W9-CBH-923

Writing Your First Play

Second Edition

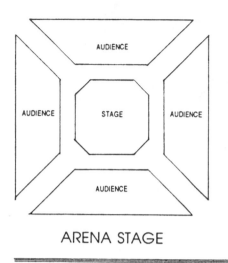

ARENA STAGE

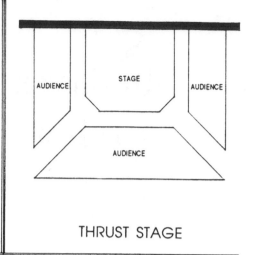

THRUST STAGE

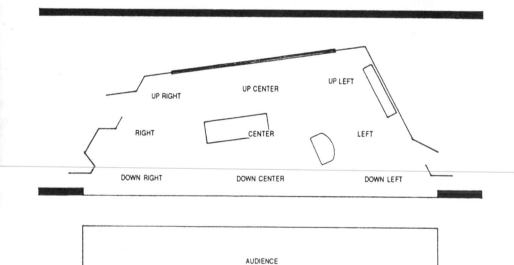

PROSCENIUM STAGE

Writing Your
First Play
Second Edition

Roger A. Hall

SOUTHERN OAKS

METROPOLITAN LIBRARY SYSTEM
SERVING OKLAHOMA COUNTY

Focal Press
Boston Oxford Johannesburg Melbourne
New Delhi Singapore

Focal Press is an imprint of Butterworth–Heinemann.

Copyright © 1998 by Butterworth–Heinemann

 A member of the Reed Elsevier group

All rights reserved.

No part of this publication may be reproduced, stored in a retrieval system, or transmitted in any form or by any means, electronic, mechanical, photocopying, recording, or otherwise, without the prior written permission of the publisher.

Recognizing the importance of preserving what has been written, Butterworth–Heinemann prints its books on acid-free paper whenever possible.

 Butterworth–Heinemann supports the efforts of American Forests and the Global ReLeaf program in its campaign for the betterment of trees, forests, and our environment.

Library of Congress Cataloging-in-Publication Data

Hall, Roger A.
  Writing your first play / Roger A. Hall.–2nd ed.
    p.   cm.
  Includes bibliographical references and index.
  ISBN 0-240-80290-X (alk. paper)
  1. Playwriting.   I. Title.
PN1661.H28   1998
808.2—dc21   97-31903
                                                  CIP

British Library Cataloguing-in-Publication Data
A catalogue record for this book is available
from the British Library.

The publisher offers special discounts on bulk orders of this book.
For information, please contact:
Manager of Special Sales
Butterworth–Heinemann
225 Wildwood Avenue
Woburn, MA 01801-2041
Tel: 781-904-2500
Fax: 781-904-2620

For information on all Focal Press publications available, contact our World Wide Web home page at: http://www.bh.com/focalpress

10 9 8 7 6 5 4 3

Printed in the United States of America

Table of Contents

Foreword

Writing a play is a snap. All you have to do is arrange a bunch of words in order so that when actors say them on stage, they will bring an audience to laughter, tears, or enlightenment. Go to it. I'll catch you later.

OK, so that's an exaggeration, but there are a slew of writing books out nowadays that do deliver the message that the important thing is to *write* the play (or novel or screenplay) and get that sense of accomplishment. Figuring out how to do so will come to you along the way.

The fact that this theory isn't applied to other professions ("The important thing is to *do* the heart/lung transplant. You'll figure it out as you go.") doesn't seem to bother anyone. Plays are emotional things, things from the heart. They aren't something you have to learn how to do.

Or do you? Faith and inspiration might get you a good scene or even a good act, but eventually you'll see that those flashes of inspiration need to be made part of a structure, that a play has to be built. In order to build something, you need to have material and tools. That's what this book provides.

I've been working as a writer for a dozen years in theater, film, and television, and I've only taken one writing course. It was the first one I ever took, and it was taught by Roger Hall. I hadn't written before then. I haven't stopped since.

I remember one exercise in particular (it's in this book) that changed my whole idea of what a play could be. Open a script and the eye clearly sees what a play is—people standing around talking, sometimes sitting around talking. So when Dr. Hall assigned Exercise 1—write a scene with no dialogue it seemed like a weird, if fun, stunt. As I wrote the scene (as I recall it was about something startlingly original, like people in dorms playing loud music), I saw that it didn't have to be strained, or forced, or twisted to fulfill the requirements. (If it was all three, that wasn't the assignment's fault.) I saw that a scene could be complete and whole without words because a play isn't about words; it's about people doing things, and speaking is only one of the things they do.

It sounds obvious. It's a simple lesson, but of the hundreds of scripts I've read since I became a producer, 99 percent of them have been nothing but people talking.

All of the exercises from Dr. Hall's class, each a giant leap forward in knowledge and experience for me, are in your hand right now. Also in it, and most entertainingly, are my fellow students. One of the most delightful aspects of this book is the writing samples from Dr. Hall's students over the years.

Most writing books, in their section on conflict, ask you to read, say, a scene from *Glengarry Glen Ross*, which leads any sensitive novice writer to give up and go into frozen produce marketing. You won't be able to write like Mamet right off the bat, and forcing a comparison like that is a recipe for writer's block. If you're serious about being a writer, you've read the great and the good plays; you know what to aspire to. Now you're trying to learn your craft.

That's a lonely process, and the other students in this book will keep you company, give you someone to compete with. Someone to make you say, "I can do that" or "I can do better than that." (Though you may be surprised; some of this writing is quite good.)

So do it. Read the book, do the exercises. Have fun, talk back. When you're through, you'll be ready to tackle that heart/lung transplant without fear of losing the audience on the operating table.

Phoef Sutton was executive producer and writer for the classic NBC television series "Cheers!" for which he won a pair of Emmy Awards. He also wrote for and produced two of Bob Newhart's endearing sitcoms. Mr. Sutton has been a recipient of a National Endowment for the Arts Playwriting Fellowship, and he was the winner of the Norman Lear Award for Comedy Playwriting and the Roberts/Shiras Playwriting Award. Most recently, he wrote the screenplays for Mrs. Winterbourne *and* The Fan.

Preface

When Focal Press first wanted to publish *Writing Your First Play*, I was elated. When people I didn't even know began to write me letters saying how much they had benefited from the book, I was even more pleased. I mean, I knew this approach worked within my classroom setting, but I wasn't entirely sure how it would translate to other teachers or to individuals laboring on their own. Apparently, it worked!

When Focal Press then asked me to prepare a second edition, I wondered to myself: What can I put in a second edition that I haven't already put in the first? I knew I could update the references to plays. I knew I could include different scenes and examples with more contemporary material. I also knew I didn't want to alter the basic structure of the book, and I didn't want to change the tone of the writing; I want that to be as personal as possible between you and me. So what difference, *fundamentally*, would a second edition make?

First, I decided to expand on one of things that people liked most about the original edition: the exercises. I've used quite a few more writing exercises with students than I actually included in the book. So for this edition I included an expanded chapter (Chapter 8) with additional exercises that illumine the art of playwriting and get your writing muscles warmed up.

Then I turned to the list of helpful suggestions that people had passed along. Those included a more substantial final play example, a fuller explanation of how to proceed from scenes to a full one-act play, and more information about what to do once the play was completed. I've tried to respond to each of those ideas by enlarging the chapter on Writing Your Play (Chapter 9) and adding a new section on Marketing Your Play (Chapter 10).

The essential core of the book, however, remains—especially the sequential exercises that help a writer learn the fundamentals of playwriting and develop a play at the same time. Just as I tell people to write from their own experience, the basic concept for that arrangement goes back to my own days as a student.

I once took a playwriting class in which the teacher, on the first day, said: "Your first assignment is to write a play. Bring it to class in two weeks, and we'll read and discuss it."

After class that day I turned to the student beside me and asked, "Do you know how to write a play?" "No," she replied, "that's why I took the class. Do you?" No. I didn't either. I realized intuitively that there had to be a better way to help people write plays. When I began teaching playwriting myself, I put my efforts together with Dr. Ralph Cohen, a colleague in the English Department, who had similar interests.

We wanted to devise an approach that would allow students to work on certain fundamental aspects of playwriting one element at a time. We also wanted to encourage plays to evolve slowly from one idea to another. After some trials and errors, we developed a series of exercises that helped students to understand and use basic elements of drama such as action, conflict, dialogue, and character.

It just so happened that our experiments with playwriting came at a good time, for the last few years have seen a burgeoning interest in playwriting. Playwriting contests have sprung up like wildflowers, sometimes in the most unlikely terrain, and many of them carry substantial financial and artistic rewards. Community theater, secondary school drama programs, and college theater departments have demonstrated a greater willingness than previously to depart from the Broadway-hit syndrome and experiment with original works. The same is true of regional professional theaters, many of which have become prominent in the development of new scripts. In addition, almost every major city has semiprofessional or professional companies that focus their attention, sometimes exclusively, on original drama.

Broadway—the New York commercial theater—which in many ways is still regarded as the pinnacle of American theater, has also responded to this surge of new plays. Because of high costs, most of the productions in the major New York theaters have hued to the economically safer road of revivals or imports of successful British plays. Even in New York, however, such original American plays as *Six Degrees of Separation* and *Three Tall Women* have flourished. In many cases scripts such as *Oleanna, The Piano Lesson,* and *Angels in America,* which were developed and produced first in other areas of the country, have eventually enjoyed Broadway success. Also, numerous off- and off-off-Broadway theaters in New York have gained a certain amount of notoriety through their production of new material.

This current fascination with the production of original scripts has led to explorations of the best and most effective ways to put a story into dramatic form. That is hardly a new concern. Ever since Aristotle some 2300 years ago tried to label the elements that comprised a superior tragedy, writers and critics have been attempting to tell people how to construct plays. Writers of our own century have not been idle in that enterprise. Dozens of volumes have been printed on playwriting, playwrighting, playmaking, how to write a play, and even how *not* to write a play. They have been written by knowledgeable critics of the drama such as Walter Kerr and Brander Matthews, experienced teachers of playwriting such as George Pierce Baker and Sam Smiley, and successful playwrights such as John Van Druten (author of *I Remember Mama* and *I Am a Camera*) and John Howard Lawson (author of *Processional*).

Every one of those dozens of volumes about playwriting contains valuable nuggets of advice—many of them reappear in the chapters of this work—and many of the books are excellent guides for the analysis of plays and the understanding of the playwright's craft. Still, as a teacher of playwriting I thought something was missing.

The development of a play is too often viewed in terms of procedure: Proceed from an idea to a brief scenario to an expanded outline to a rough draft of the play. Sometimes the development of a play is seen in terms of a progression of analytical elements: progress from an inciting incident through a series of minor conflicts or crises to a major crisis (the climax) and thus to the conclusion or denouement. Both of those standard methods seem to assume that the writer has the ability to invent a relatively complete idea for a play and needs only some working procedures or the inclusion of various standard dramatic devices in order to bring the idea to fruition.

My teaching experience suggested otherwise. I saw that young men and women who had never before written a play had much more basic needs. First, they needed some experience in working with the fundamental building blocks that make a play a play—such things as action, conflict, and the interaction of characters. Second, they needed help in generating an idea for a play. The concept of "getting an idea" for a play is a complicated notion. Ideas for plays seldom emerge fully grown in the playwright's mind. More frequently they take root as a seedling. They grow and develop bit by bit, with the introduction of a scene here, a new character there, and an idea or two over there. Eventually the idea progresses and the sapling takes on the stature and the shape of a mature tree with trunk, limbs, and branches. The foliage is brilliant with vivid characters, dramatic action, and snappy dialogue.

Just as it is hard to imagine how a mature tree will grow from a seedling, it is extremely difficult for a beginning writer to envision a fully developed play. It is also almost impossible for an inexperienced writer to manage with any degree of skill the many essentials of a play—from action, conflict, and environment to dialogue and characters—all at one time. It would be as if a music teacher were to tell a young student who was learning composition, "I want you to compose a symphony"—without first instructing the student in harmony, progression, syncopation, and the other necessary musical elements. Complex creations such as symphonies, plays, and trees unfold in stages. And those who are learning to create them need to practice with one or only a few elements at a time before combining them. The approach I suggest allows plays to develop. Through the sequence of exercises I describe, scenes become progressively longer, more complex, and more sophisticated until, like leaves in spring, plays appear.

This book is designed to help the beginning and inexperienced playwright. It presents for students a series of exercises to provide practice in working with the essential ingredients of a play. Because the exercises are sequential, one leads naturally to another, becoming more complete and more complex with each step. The exercises provide an opportunity to begin with a seedling of an idea and nurture it carefully before asking it to bear the fruit of a mature tree.

This approach has been quite successful with the undergraduate students I teach. In the almost 20 years that I've been working with original scripts, James Madison University has produced over 75 student-written plays including one-acts, full-length pieces, and musicals. Most of those were mounted in our Experimental Theater, but several were a part of our main season of plays. Ten of the plays were recommended for competition by the American College Theater Festival, and five received awards in various writing competitions.

Some of the students who have used these exercises are still writing successfully for the stage, for the screen, and for television. Many have gone into professional theater in other capacities, as managers or performers. Several are writing for newspapers or magazines, and others are teachers, including some who are using these exercises with *their* students.

In addition to the work I've done with my own students, I also shared these exercises with English and drama teachers at the college and high school level, and they have reported using them successfully with their students. The exercises also formed the basis of a playwriting unit for high school students at a young writers' workshop at the University of Virginia.

Of course, for several years now the series of exercises has also been available for writers through the first edition of this book. In fact, one of the most gratifying aspects of the publication of the book turns out to be the letters and responses from writers and teachers who read it, like it, use it, and are kind enough to send along their positive comments.

In this book I make numerous references to a variety of plays and movies—classics and modern—from Sophocles and Shakespeare to Eric Bogosian and *The Sisters Rosensweig*. In many cases, however, I've used hypothetical examples or short scenes written by students. I've done that for several reasons. Although great writers can provide marvelous models, their works are often longer and more complex than I want for the illustration of particular points. Also, they do not reflect what someone using these exercises might actually write.

The student writings, on the other hand, provide concise examples of exactly what the exercises can be expected to produce. They are also relatively short, and they are finished. That is, even if they suggest a continuation, they represent a whole response to an exercise rather than a scene taken from a longer play. For that reason they are simpler to work with than a scene that is part of a more complex project.

Many people contributed to this book. I would like especially to thank Dr. Ralph Alan Cohen of the English department who worked with me to originate this sequential approach to playwriting. I have appreciated the opportunity to try out my ideas at James Madison University, and in particular I want to express my gratitude to my colleagues in the theater program who encouraged the efforts of student playwrights. I would also like to thank the many people who provided commentary on the first edition, and, in particular, Ramon Delgado of Montclair State University and Barbra Graber of Eastern Mennonite University; Bob Small of the ShenanArts Playwrights Retreat; and playwrights Stephen Gregg and Bill Rough for their positive remarks and helpful suggestions. Finally and most importantly, I want to thank my playwriting students who have consistently amazed, surprised, and entertained me with their work and who generously permitted me to use examples of that work in this book.

Introduction

The direct and primary goal of this book is to enable a person to write a play of at least one act in length. A one-act play is usually, though not always, a piece of continuous action in one setting about 30 to 40 minutes in playing time. That, however, is merely a guideline. Some one-act plays are as short as five or ten minutes. Others use two or more different scenes or units of actions, while still others use locations that shift from one place to another.

The purposes of this book are all relatively simple. If you are interested in writing a play, this book provides a series of exercises designed in sequence to help you do that. You should read the assignment, the explanation, the examples, and the evaluations. Then you should do the assignment. Then compare your writing with the examples and look at it in light of the commentary on those examples. The one important part missing, of course, is someone with an artistic bent to provide an outside opinion. If you are using this guide on your own, you will have to provide your own objectivity.

Another purpose is to enable a teacher to provide a basic course or unit in playwriting. Most teachers of theater have the intelligence and artistic sensibility to provide helpful comments on inexperienced writers' first attempts at playwriting. Many, however, are unsure about what assignments to use in the actual development of a play. The step-by-step orientation of this book is designed to solve that problem.

Not everyone who sets out to write a great play will be successful, but everyone who attempts the exercises here will come away with three very important, indirect benefits. First, a writer experimenting with dramatic formats will gain a greater appreciation of plays, of the ways in which they are constructed, and of the choices the playwright makes. Second, the exposure to working in a dramatic mode will help a writer in other formats; the emphasis on action, character, conflict, and dialogue will affect the way a writer thinks, even the writer who is later working in poetry or prose. Third, the writer will gain a greater awareness of human behavior, of the intricacies of personal relationships, of the reasons why people do things to each other, and of the ways they react to their environments.

People reacting to their environments: That's one of the drama's primary concerns, and it's one of the reasons why the first section deals with "action."

1

Action

EXERCISE 1 • DESCRIBE AN ACTION
Write a description of an action taking place. Describe only those things that can be seen. Use no dialogue, although you may use other sounds. The scene should take place in one location, but don't worry about stage terminology. Minimum length: 1–2 pages.

Drama is the imitation of an action. Unfortunately, most of us experience plays by reading them rather than by seeing them. It seems hard to believe, but countless individuals have never seen a live play performed. Even for those of us who have seen scores of plays, we've undoubtedly read more than we've seen.

When we read a play, we see characters' names and the lines they speak. Very little of what the characters *do* is described. That's particularly true of classical plays. Shakespeare's plays, for example, which are often an individual's first exposure to drama, have little in the way of stage direction or descriptions of particular actions. Rather, in those plays most of the action must be deduced from the spoken line.

When Macbeth returns from killing King Duncan, only a reading of one line of dialogue three-fourths of the way into the scene informs us that Macbeth has thoughtlessly brought the murder weapons with him instead of leaving them behind. In performance, of course, the action would provide that information immediately. We would see the knives in Macbeth's hands as soon as he entered, and we would continue to see him holding them throughout the scene.

The fact that most of us receive our introduction to plays through print rather than performance has serious consequences for beginning playwrights. Since for the most part we only read the lines spoken by characters, is it any wonder that we come to think of plays as dialogue between characters in which the words are all-important and tell us everything we need to know? And is it any wonder that when beginners sit down to compose a play, what usually develops is a group of characters sitting around talking?

Our first task, then, is to create a different perspective on what constitutes a play. We must find a way to present drama in terms of actions rather than words, in terms of what people *do* rather than what they *say*.

Action is one of the building blocks of drama. Drama is about what people *do*, and what people *do* is *action*, so it is worthwhile to look at the idea of action in some detail.

Action is when someone does something. But the concept of action is really much more complex and multilayered. Let's look at an example: A woman walks into a fast-food restaurant. She stops and looks up at the items available and their prices. She gets into line. She pulls her wallet from her purse and takes out money. She reaches the front of the line and requests, "A cheeseburger, small fries, and a small orange." She pays for the food, takes it, walks to a table, and begins to eat her meal.

There are numerous actions taking place in this brief scene. Walking into the restaurant is one action. Reading the menu is a second. If the woman puts on a pair of glasses while she's reading, that's a third. And so on. It quickly becomes clear that not all her actions are equally important. Walking in is probably not as important as ordering the food, or as eating it. Hence, even the simplest actions have a hierarchy of importance with respect to other actions. Playwrights as well as directors and actors must understand how actions fit together and decide which actions are most significant.

While not all actions are equally important, all actions can be important in different ways. I suggested that walking into the restaurant was not as important as ordering or eating the food. Nevertheless, walking in is essential in that it gets the woman into the restaurant. The rest of the scene can't take place without it. In some instances the entrance of a person might be the most significant action. The woman, for example, might hurry into the restaurant and then look carefully behind her out the window.

Reading the menu constitutes another action. Is it important? In one sense it isn't. It could be eliminated, and the woman could still get her meal. But that action provides revealing information. That the woman looks at the menu might indicate that she doesn't frequent the restaurant, that she's indecisive, or that she's concerned about how much things cost. Any of those elements might be extremely significant. Consider how different the scene would be if the woman simply entered, walked up to the counter, and ordered.

In both cases, whether she reads the menu first or not, the woman gets her meal, and here is where actions become multilayered. This scene contains many actions. Taken together, however, they constitute a larger action: A woman buys her lunch. If we also saw her eating her meal and leaving, the whole action might be "a woman lunches."

The same layering of actions occurs in any play. In one moment Hamlet praises an actor. That is one action. A series of actions taken together might constitute a larger action: Hamlet sets a trap. All of those actions taken together form an action that absorbs the character throughout the play: Hamlet avenges his father's murder. In acting terms, that overall action is what is meant by such terms as *through line, superobjective,* or *spine.*

The recognition that small actions fit together to create larger actions is just as important for a playwright as it is for an actor.

Taken simply, an action is someone doing something. The action can be small in scope or large; it can be a simple action or a complex action composed of numerous smaller actions. It can be static action—such as sitting or sleeping—or frenetic action—such as running or dancing.

Most modern theater practitioners also recognize psychological action—a thought process, a decision, or a point of view. For instance, if a lawyer tells you to do something, his authority might compel you to do it without even the slightest actual physical action on his part. There is a psychological force at work.

Harold Pinter is famous for the silences written into his scripts. Pinter knows that silences represent critical moments because people make decisions in those pauses and silences. We, however, cannot *see* the decision. What we do see is the physical manifestation of the decision: The character does something or says something based on her decision.

Working with beginning playwrights, I've found it especially helpful to consider thought and decision as preceding physical action. Picture, for example, a woman in a department store. She looks at a bracelet. Then she looks around. Thought occurs. A decision is reached. The woman taps a bell on the counter to call a salesperson.

At the moment the woman is looking around, we may think she is considering pocketing the bracelet. The action that followed—tapping the bell—reveals the thought process that preceded it.

At this beginning stage, we want to focus on basic action—such as the woman ordering her lunch—that is large enough to comprise several other actions. It is sufficient for now that we understand that the action may in turn be part of a larger, more complex whole.

DRAMATIC ACTION PROVOKES QUESTIONS

Writers understand that action by itself—no matter how many cars you blow up—does not guarantee interest. Now for some circular thinking. For an action to gain the interest of the audience, it must be dramatic action. But what is dramatic action? And how is it differentiated from other action? Dramatic action is action that gains the interest of an audience.

William Archer, a British dramatist and critic and the man who translated Henrik Ibsen for English-speaking audiences, wrote nearly the same thing. His turn-of-the-century study of playwriting, which still represents sound thinking, defined *dramatic* as "any representation of imaginary personages which is capable of interesting an average audience assembled in the theater."

The first time I read that, I regarded it as totally inadequate. Surely, I thought, we should be able to determine what is dramatic within the context of the script itself. But Archer realized, as I did only somewhat later, that the audience is essential, and a play must undergo the test of an audience or experienced readers to determine its dramatic content.

Dramatic action is realized only when what the author *thinks* is dramatic is effectively conveyed to others.

A perceptive playwright can gather some notion of what is likely to be dramatic activity. To explore that, let's look at dramatic action from several different angles. One approach is to see dramatic action as action based on the desire of a human being to attain a goal. A man walks onto a stage with a painting and an easel and begins to set up the easel. It falls over. He tries again. It falls again. He carefully inspects the easel, looking at the legs and hinges. He tries yet again to set it up.

In this instance we understand the goal—to set up the easel. Perhaps we infer the larger action—to display the painting. As the man tries to accomplish his goal another element of dramatic action comes into play—namely, tension or conflict. The human will to attain the goal is thwarted by the obstacle of the inanimate object. The man is in conflict with the easel. As a result of the conflict, tension arises.

The human will to attain a goal. Conflict. Tension. All of those elements work together to induce *questions*. Will the man succeed in setting up the easel? What if he doesn't? What will he do then? What does he want to do with the painting?

The scene could continue in a variety of ways. The easel falls again. The man picks it up and bashes it to pieces. Then he picks up the remains of the painting and calmly walks off the stage. An alternative: The easel falls again. The man goes offstage, comes back with a chair, sets that on the stage, and displays the painting on the chair. Another alternative: The easel stays up. The man sets the painting on the easel. In all cases the questions have been answered and the action is ended.

Ingmar Bergman, the Swedish director and dramatist, once commented on how he formulates ideas. Bergman explained that he literally has visions. He might visualize people in a pink room. One might be turned away, looking out a window. Bergman would ask himself questions about his visions. Why was the one woman turned away? What was outside the window? Why was the room pink? What was going on between the two women? If the answers to his questions were interesting to him, he'd ask more questions until, perhaps, he had something to write. Essentially, said Bergman, he wrote not because he had something to say but because he had questions to ask.

Jon Jory, the producing director of Actors Theater of Louisville and one of the primary forces in the development of new playwrights in the United States, has said that a play should always pose a question within its first five lines. Two or more questions, he noted, would be even better.

But, significantly, the questions need not come in the form of dialogue. In fact, in good drama the questions are usually inherent in the *action*, and the dialogue—questions expressed verbally—simply adds to the questions already raised implicitly by what we see occurring.

The opening scene of Arthur Miller's *Death of a Salesman* illustrates how dramatic action provokes questions. Willy Loman, the 60-year-old salesman, enters carrying two large sample cases. He crosses to the doorway of his dimly lit house. He unlocks the door, comes into the kitchen, and sets the cases down. He feels the soreness of his palms and a word-sigh escapes his lips. He closes the door and carries the cases into the unseen living room. We see Linda, his wife, stir in her bed. She listens to Willy's entrance, then gets up and puts on a robe. The dialogue begins:

Linda: Willy!

Willy: It's all right. I came back.

Linda: Why? What happened? (Slight pause) Did something happen, Willy?

Willy: No, nothing happened.

Linda: You didn't smash the car, did you?

Willy (with casual irritation): I said nothing happened. Didn't you hear me?

Linda: Don't you feel well?

Willy: I'm tired to the death.*

We *see* the action as a tired old man enters his house late at night. We *see* a woman obviously concerned about the man. He wants to rest. She wants to know what's going on. Each person wants to attain certain goals. We *see* the man dismiss her concern. The human wills are in conflict. We *see* a relationship straining. Tension.

This excellent opening scene induces numerous questions. Why is he coming in so late? Why is he tired? Why is she concerned? Those are merely three questions out of many. And all of that has been provoked by the simple action that occurs even before a line has been spoken.

Miller then uses a variety of questions in his first few lines to focus our questions. Some of them are significant. "What happened?" asked twice tells us that Willy wouldn't be there if something weren't wrong. The question about the car informs us he has the capability of smashing it, as he does at the end of the play.

There are significant elements that are *not* present in my summary of this opening scene. In my condensation of Miller's opening stage directions I omitted, among other things, the following reference to Willy: ". . . his exhaustion is apparent." I omitted it because I wanted you, the reader, to make that conclusion based on Willy's *actions:* feeling his sore palms and sighing.

DRAMATIC ACTION EMPLOYS VERBS
Just as Willy's exhaustion is exposed through his actions, so dramatic action is expressed in verbs, in what people *do*, not in moods. It is

---

*From *Death of a Salesman* by Arthur Miller. Copyright 1949, renewed copyright © 1977 by Arthur Miller. Reprinted by permission of the publisher, Viking Penguin, a division of Penguin Books USA Inc.

extremely difficult for an actor to play a general mood such as anger, and it is well nigh impossible for a playwright to write an effective general mood. The dramatic mood arises from the specific actions undertaken.

When we say to ourselves, "That person is angry" or "That man is exhausted," we are making a conclusion based on certain actions we have seen performed. We do this so frequently and so automatically that we hardly even think about the actions we are seeing that prompt us to leap to the conclusions. We just leap.

Test this. The next time someone describes someone else as "angry" or "happy" or "sick," ask that person why he or she thinks that. The conversation is likely to run something like this:

```
Friend: Did you see Frank today? He really looked
        awful.

You:  Oh yeah? How come?

Friend: Well he didn't look like he felt very well.

You:  Why do you say that?

Friend: I don't know. He just looked sick.
```

There it is. One conclusion ("He looked sick") used to justify another conclusion ("He didn't look like he felt very well") to justify another conclusion ("He really looked awful").

Perhaps the most startling statement made by your fictional friend is: "I don't know." Of course your friend knows. The friend must know, or else how could your friend have arrived at the conclusion that Frank is sick. Perhaps Frank was moving more slowly than usual, or was hunched over. Perhaps he coughed, sneezed, grimaced, groaned, or limped. All of those are *actions*, and whatever the symptoms, your friend saw those actions. But instead of registering: "Frank is coughing and talking hoarsely," your friend automatically registered the general conclusion: "Frank is sick."

The problem with that normal and generally quite effective way of proceeding to conclusions is that many times the conclusions can't be visualized, and playwrights must deal with what can be seen. Just as Pinter's silent decisions cannot be seen but are revealed by the actions that follow them, so, too, a general mood is discovered by the actions that create the mood.

Let's look at another example of the difference between a general mood, which cannot be seen, and a specific action, which can be seen and which establishes the necessary mood. A hypothetical stage direction reads: "Jennifer enters the room. She is angry." That's a problem. Nothing is visible. We have no outward manifestation (action) to suggest the inner emotion. Instead, the stage direction might read: "Jennifer hurries into the room. She stops and looks around. Her eyes land on the stereo system, and she moves to it. She quickly looks through several compact discs and then selects one. She yanks the disc out of its jewelbox, throws the jewelbox to the floor, and smashes it with her foot." That is a dramatic action. We *see* the action. We understand the mood. And we are provoked to ask questions about what is going on. We might wonder not only why Jennifer is so upset, but also why that one particular CD was so significant?

This notion of looking for the action that prompts conclusions rather than settling for the conclusion is equally important to performers. Playwrights occasionally slip, and when they write "Jennifer is angry," it's the performer who has to translate that generality into concrete action that is both quintessentially human and wildly interesting.

Many individuals turn to writing because they think they have something to say. In writing plays, that can be a problem. I have a rule that I repeat over and over again to beginning playwrights: DON'T SAY IT, SHOW IT!

Thornton Wilder, in his superb essay on playwriting, expressed it this way: "A play is what takes place. A novel is what one person tells us took place." So if you have something to *tell* us, write a novel. If you have something to *show* us, write a play.

With that as a reminder, let's proceed to the first exercise.

• • • EXERCISE 1.1
Write a description of an action taking place. Describe only those things that can be seen. Use no dialogue, although you may use other sounds. The scene should take place in one location, but don't worry about stage terminology. Minimum length: 1–2 pages.

Below are two very basic examples—one acceptable, one unacceptable.

EXAMPLE

```
It's dark. A man lurks in the bushes beside a house.
He looks around, then slides carefully to a window. He
looks around again, then tries to lift the window. It
doesn't move. He takes a small tool from his pocket
and scores a square on the window. He hits it with a
gloved hand, and a section of the window drops out. He
reaches his arm inside the window. A light comes on in
the room. The man quickly withdraws his arm, turns, and
runs from the house.
```

EXAMPLE

```
It's a dark room. A man and a woman are sleeping. The
alarm clock goes off. The man switches off the alarm.
It says 6 A.M. The man takes a gun from beneath his
pillow, looks at it, then places it beside the clock.
The man continues to sit, groggily thinking of the
tequila he guzzled the night before and the short rest
he has had. He rises and glances at a paper on the
```

table. A headline reads "Bank Heist Sets Record." He goes to the closet where he takes out pants and a shirt. He begins to put on the shirt and can't button the sleeve because a button is missing. He strides angrily to the bed and wakes the woman. He holds out the buttonless cuff. She shrugs and turns over. The man finishes putting on the shirt as best he can. He puts on his pants, finds his socks and shoes, puts them on, and exits.

EVALUATION

As you read the scene, can you picture the action as it occurs? Is anything missing or unclear? Would you, for example, be able to see a headline or a small clock on a bedside table? Probably not on a stage. Even with a movie close-up, you might not know if it were 6 A.M. or 6 P.M.

I indicated that questions are important to dramatic action. Questions raised can also be a problem for a play, especially if there are too many questions or if significant questions are left unanswered. Examine what questions are raised by the action and whether they are answered. In the first example, a key question is: "Will the man succeed in breaking in?" While we may not know why he wanted to get in the house, that basic question, at least, is answered. In the second example we see a man getting dressed and upbraiding the woman in the room. But what about the gun, which the man places prominently on the clock and never returns to? What's it for? And the bank heist? Did he do it? Who knows. The author tries to include too much, and the information overload becomes confusing.

Next ask yourself: "Does the example contain anything that can't be seen?" The second example has numerous problems in this area. There is the difficulty of the time, for one thing; the newspaper presents a similar problem. But more importantly, we can't read characters' thoughts except through their actions, so we can't possibly know the man is thinking of tequila or his short rest.

The novelist or the short story writer can take us directly into the mind of a character. If the novelist writes, "Turk sat groggily on the bed, thinking of the short rest he'd had and the tequila he'd guzzled the night before," then that's what is on Turk's mind.

The playwright does not enjoy that option. Even if a character speaks directly to the audience—as Iago does in explaining his actions in Shakespeare's *Othello*—the character may or may not be telling the truth. Or the character may not even fully understand his or her own motivations.

The playwright must find means to make visible to the audience the outward manifestation of the characters' thoughts. For example, the playwright might write: "The man rises unsteadily and walks slowly to the closet. He glances at an empty bottle of tequila lying on its side on a table. He stops, picks up the bottle, and holds it upside down as he watches the last drops fall on the table. He pitches the empty bottle in a trash can and proceeds to the closet." That would clarify the relationship between the man and the bottle.

There are moments in plays when we, as the audience, feel as though we know exactly what the character is thinking. In those moments the work of the playwright and the performer come together in such an effective way that the external activity illumines the inner thought process. That activity can be as subtle as a look or a shrug or as overt as a statement or a punch. In any case, the playwright and the performer are carrying us along, and we understand why characters are behaving as they do.

Are there places where the scene seems to need words? In the first example there are none, although voices might be heard when the light comes on. In the second example the moment between the man and the woman seems to want dialogue. In that instance lines could add something to the action.

There is another question: "Does the scene suggest a mood?" The mood of the first one is definitely melodramatic. You can almost hear suspicious-sounding music underscoring it. The mood of the second piece is indistinct. The style is realistic, but what is described could be played comically or despairingly.

A final question about the descriptions is: "What kind of characters are expressed through the actions?" In the first example the man in the bushes seems calm, orderly, and experienced. In the second example the man's situation seems almost out of his control. The woman in the bed certainly pays no heed to his remonstrations.

Put your skills of evaluation to work on one final example, this one written by an undergraduate student.

EXAMPLE
A MAN IN A BUS TERMINAL
By Lynda Edwards

A tiny middle-aged man wearing a rumpled suit is sitting in a big-city bus terminal. It is night. The clock on the wall says 11:45. He has a small overnight bag with him. The man glances frequently at his watch and at the clock on the wall, and at two disheveled bums sleeping on the floor near the doorway. A muscular man wearing a T-shirt proclaiming "Olympic Sex Team" under a leather jacket enters and flops down next to the man. The man nervously unfolds a newspaper and tries to read as the Leather Man glances at the paper over his shoulder. The paper gets tangled and crumpled as the tiny man desperately tries to smooth it. Finally he smiles hesitantly at the Leather Man, wads the paper into a ball, stuffs it under his arm, and lunges toward the ticket counter with his coat and bag.

He pulls out his ticket from his pocket and slaps it on the counter. The attendant behind the counter slowly walks over, eating a sweet roll. The small man slaps the ticket angrily and points to the clock behind the counter. The attendant holds the roll in his mouth—crumbs and powdered sugar falling to the counter—as he inspects the ticket. Then he glances at the clock, shrugs, hands the ticket back to the man, and walks away.

The man stuffs the ticket back into his pocket and goes to the bathroom, which is a small, windowless room. He slams the door and turns the lock vigorously. He splashes water on his face, sighs, takes a deep breath. He smiles, looks at himself in the mirror, straightens his shoulders, and reaches for a paper towel. There are none. He shakes his hands as he marches for the door. He cannot get it open. He yanks and tugs. He puts one foot on the door and pulls. He backs up, laughs, and shrugs. Then he whirls and attacks the door, tugging hysterically. He calms. He knocks timidly, then louder. He pounds on the door with both fists. He kicks the door. He looks around the bathroom. He goes to the sink and suds up the soap. He splashes the wet liquid on the doorknob and lock mechanism. He tries again to open the door, but his hands just slip. He looks at his wet hands. He reaches again for paper towels that aren't there, and then hits the metal container. He dries his hands and the door handle with his coat. He goes to the corner of the bathroom and sits down with his bag. He takes a package of Rolaids out of his bag and pops one into his mouth. He makes a pillow out of his coat and puts it behind his head. He leans back, curls up in the corner, and closes his eyes.

THE END

EVALUATION

This lighthearted piece answers all the questions it raises. The author uses virtually every small detail that she establishes. The man's suit becomes both his towel and his pillow, and the Rolaids come from the overnight bag. Furthermore, everything can be seen. Even the clock in the bus terminal would be large enough, and if it were dark out and the lights in the terminal were on, it would be apparent that the time was evening.

Although everything in the scene is clear without words, dialogue might add to the enjoyment of the piece in two or three places. We might enjoy the conversation between the small man and the messy attendant. And wouldn't the small man call out in his attempts to get out of the bathroom? A few words between the Leather Man and the small man might also accentuate the threat our hero feels.

The important thing to remember is that in those instances lines could add to the action. Thus we can understand dialogue as being a complement or an accessory to the action. Words, rather than being an end in themselves, become merely one method that people employ to accomplish certain aims or perform certain actions.

The scene is obviously comic in mood, as would be established in the meeting of the small man and the imposing (though apparently harmless) Leather Man. As for character, the scene presents a timid unfortunate soul not dissimilar from some of the great silent movie heroes. You can almost visualize Charlie Chaplin, Buster Keaton, or Harold Lloyd confronting the Leather Man, the attendant, or the locked door.

WHERE IDEAS COME FROM

As you prepare to write your own "silent scene," you might wonder where the ideas for the scenes come from. They can come from anywhere—from your observations or from your imagination. If you want a starting mechanism, I suggest you think of a person in a particular environment, something as simple as a woman on a beach looking around. Then, like Ingmar Bergman, ask yourself questions. What is she looking for? Something she lost? Someone she knows or is waiting for? A bag? A key? A missing child? When you ask yourself questions, situations and actions immediately arise.

Another prod to the imagination is to find something incongruent or out of place in a situation—like one of those "what's wrong with this picture?" puzzles. Suppose that the woman on the beach is wearing an elegant gown. Or perhaps she's carrying a common but unbeachlike article such as a briefcase. Again, the questions will lead you to a situation and an action. If they don't, try a different person in a different environment.

Writing an action is a major step for a playwright because it establishes a particular viewpoint toward drama. It demands a mind-set geared toward seeing a play in terms of action, in terms of people doing things to each other and interacting with each other. That concept of drama as action will remain the basis for the following exercises. Therefore, as we move on to more complex exercises integrating various dramatic elements, always keep one question foremost: "What do we see these people doing?"

2

· · · · · · · ·

Direct Conflict

EXERCISE 2.1 • DIRECT CONFLICT
Place two characters in a scene of direct conflict. Write the
scene in stage terms indicating how the scene should be staged.
Use dialogue as needed, and resolve the scene. Minimum
length: 2–3 pages.

This chapter asks you to compose a scene that places two characters
in a situation of direct conflict. The exercise encourages you to conceive a
dramatic scene in terms of conflict. Taken in combination with the first
exercise, it attempts to substitute for the notion of people sitting around
talking the more dramatically functional notion of people doing things that
are in conflict.

The idea that conflict is essential to drama is hardly a new one.
Ferdinand Bruntiere expressed it at the turn of the century as an extension
of the nineteenth-century dialectical theories of Friedrich Hegel. It has
remained a tenet of dramatic criticism ever since. More importantly, the
earliest dramatists instinctively knew that conflict leads to good drama.
Antigone confronts Creon. King Lear's daughters reject him.

The nineteenth and twentieth centuries saw various attempts to
produce drama without traditional conflict. Naturalistic "slice-of-life"
drama was an attempt to study human beings in a scientific manner, to
examine daily lives without resorting to contrived conflicts.

In the world of the 1990s, music videos, avant garde, and even
more traditional theatrical presentations often juxtapose words, sounds, and
visual images, which feature little conventional conflict, in an effort to
induce emotional responses. In the popular musical *Quilters*, which deals
with the lives of frontier women, most of the women confront personal or
external obstacles rather than typical opposition figures. Similarly, Anna
Deavere Smith, in her riveting one-woman shows *Fires in the Mirror: Crown
Heights, Brooklyn, and Other Identities*, and *Twilight: Los Angeles, 1992*, pro-
filed numerous characters facing personal demons and obstacles even as the
plays explored the extreme violence of the Brooklyn and Los Angeles riots.
The 21st century is likely to see more "montage drama," which will seek to
elicit emotional reactions through the manipulation and layering of various
aural and visual stimuli.

Despite the emergence of those and other "nonconflict" dramatic modes, conflict has been and remains such a fundamental ingredient of storytelling that a beginning playwright needs to confront conflict head on.

The desires of different characters come into conflict within plays as a whole and within individual scenes within those plays. Sometimes the conflict is psychological—for example, when Hamlet considers his choices or bandies with Polonius; sometimes it is physical—when Hamlet duels with Laertes. Sometimes the conflict is subtle—when Macbeth calculates with his wife; and sometimes it is overt—when Macbeth finally meets Macduff. Whatever its shape, conflict provides an important starting place for a beginning playwright.

The notion that conflict is essential to drama bothers some people. They deplore the idea that plays must be about people fighting or competing. A more inclusive—less "martial"—term would be *obstacles,* for characters who lack obstacles and challenges are intrinsically undramatic. If a woman wants to climb a mountain, and she does so with little effort and no difficulties, there is no dramatic interest. If, on the other hand, she must overcome debilitating illness, harsh weather, equipment failures, and exceptionally difficult terrain, those obstacles challenge her will to succeed and heighten the drama of her quest. Viewed in that context, conflict, as we will see, is just one form of the obstacles characters must face.

OBSTACLES AND CONFLICT

Obstacles and conflicts do not arise out of nothing. They are intrinsically tied to two other components: *intention* and *outcome.* For obstacles and conflict to exist, there must be an intention or a goal. There must be a character who wants something and who undertakes action to achieve that intention. If someone or something prevents or hinders the immediate attainment of that goal, then conflict exists—just as it does between the woman mountain climber and her environment. Eventually some resolution to the conflict occurs, and that is the outcome.

The hindrance—the obstacle or the conflict—can arise in three ways. It can be internal, such as a physical or mental condition within the character. A boxer growing old is in conflict with the internal aging process of his own body as well as with his external opponent. A young artist trying to decide if she should risk making it or settle for a more secure future is battling internal psychological demons.

A second type of obstacle or conflict arises from circumstances outside the character. The mountaineer faces a climb up a sheer wall. A hiker lost in the desert faces heat, thirst, and vast distances. Both the mountain and the desert present challenging circumstances of the environment external to the character.

A third type of obstacle or conflict presents itself when the intention of one character is at odds with the intention of another character. The mountaineer wants to proceed, but her companion refuses. A young boy wants to join the army, but his mother opposes his desire. Good drama typically combines internal obstacles, external obstacles, and human conflicts. Lear, for example, battles the storm, his pride, and his approaching madness as well as his daughters' ingratitude.

The exercise in this chapter concentrates on conflict between opposing parties because that is generally the most frequently used obstacle. Because such conflict deals with at least two people, it easily reveals contrasts of human nature. In fact, many instances of internal and circumstantial obstacles are elaborated in terms of conflicts with other characters. Hamlet's indecision—an internal conflict—is conveyed not only in monologues, but through a series of conflicts with Ophelia, Polonius, Gertrude, and other characters. Willy Loman's internal problems of guilt, aging, and lack of self-respect are dramatized through external conflicts with his wife and his sons.

OUTCOMES

The *intention* of a character meeting some form of *resistance* produces conflict. To those ingredients must be added another—*outcome*. There must be some resolution to the conflict. In most plays, taken as a whole, the main character either succeeds or fails. A character overcomes obstacles and the opposing forces or is overcome by them. Our mountaineer either makes it to the summit of her mountain or she does not.

There are, however, resolutions other than complete victory or total defeat. In *No Exit* the central characters will never achieve their goals but are destined to pursue them forever in a hellish eternal conflict. Likewise in *Waiting for Godot* the dream of the characters—the arrival of Godot—will never be realized even though the characters will continue their determined waiting. In both cases the situation is resolved or ended through the recognition by the audience—though *not* the characters—that the goal will prove eternally elusive.

Scenes of a play have the same characteristics of obstacles and conflict that whole plays have. Characters pursue different intentions, they come into conflict, and the conflict reaches an outcome. But whereas a scene is one continuous bit of action in a single location, a play may include numerous locations and times. So one scene from a larger work cannot show the entire story. A scene represents one incident in the larger whole, and, although it includes a part of the story, it also leaves important elements unresolved. The outcome of a scene—except the last scene of a play—is not the final resolution or outcome of the conflict. I should add, however, that some plays—even some rather lengthy plays—tell their whole story as one continuous action within a single scene.

Within a scene, the outcome proceeds in one of three ways: 1) one character wins and the other loses; 2) the characters compromise; or 3) the characters reinforce their disagreement. But the scene may be only one segment of a much larger conflict. Hence, Hamlet in one scene persuades the players to perform a certain play. He achieves his objective. But his victory is merely one part of a larger intention, and in later scenes he pursues other objectives within that larger intention.

In the recent popular movie *Jerry Maguire,* Jerry and Dorothy have an argument that nearly ends their marriage. The incident is resolved when they go their separate ways. The scene reinforces their conflict. Neither character, however, is pleased with that outcome, so the conflict is resumed and concluded in later scenes through a process of accommodation and reconciliation. It is important to remember that the action of a scene can be resolved or ended even though the basic conflict remains.

I have chosen the word *outcome* to indicate that a conflict of intentions may be ended or continued. Other writers have used other terms, and two of those terms—*crisis* and *climax*—have been used so frequently that they require comment. Unfortunately almost every author who has used those terms seems obliged to define them somewhat differently from anyone else. Everyday language uses *crisis* to indicate virtually any kind of conflict. Hence a person trying to make a decision (internal conflict), a person trying to climb over a wall (external circumstance), or a person battling another person (conflict of intentions) all present a scene or moment of crisis. *Climax* has generally been used to designate that point of the conflict when one side wins or loses decisively or when another form of resolution (such as a compromise or reconciliation) is reached.

The exercise for this chapter asks you to put two characters in a direct conflict. You will notice almost immediately that virtually any two characters can be placed in such a conflict situation. Test it yourself. Think of any two people or characters. Then devise some situation where they can come into conflict. There's *always* a way.

To arrange people in conflict, think first of any two people. Next, think of a place or an environment where the two of them might come in contact with each other. For instance, if one of your people were a taxi driver, the location might be a taxi. In one were a lawyer, the location might be a courtroom or a law office. Alternatively, you might select a public place where *any* two people could come in contact with each other: an airport lounge, a baseball game, a lunch counter, or a street corner.

Once you have the two characters in the same place, you need to generate a conflict. That might have something to do with the occupation of one of the characters, such as the route the cab driver took or the lawyer's legal strategy. In many cases conflicts involve objects. Material items have a way of focusing disagreements. Perhaps the passenger in the taxi wants to eat some particularly offensive or messy food in the car, or wants to transport an item that the driver doesn't want in the vehicle.

STAGE TERMINOLOGY

In this exercise you will be working with stage terminology and stage directions for the first time. Virtually all plays describe a setting for the action, an environment where the scene takes places. George Bernard Shaw liked to write lengthy, entertaining descriptions that were practically comic essays. Earlier in this century Shaw's style was much admired, but few writers possess Shaw's intellect or his comic skills, and such discourses are now rare.

Shakespeare, of course, almost never described a setting except in the speeches of his characters. Some contemporary writers provide little more than Shakespeare. David Mamet's entire description of the set for *American Buffalo,* which was an elaborate, realistic second-hand store with hundreds of items, read simply: "Don's Resale Shop. A junk shop."

Your own description of a setting should probably be somewhere between Shaw's expansiveness and Mamet's terseness. You need not describe the setting in great detail, but you should mention items that are needed for the action of the piece. You should also include selected elements that establish the proper mood or succinctly define the characters who inhabit the environment.

Stage directions are normally written for a proscenium stage configuration. A *proscenium stage* is one of the three kinds of formal stages. In an *arena,* or theater-in-the-round, the audience surrounds the action on all four sides. A *thrust stage* juts out into the audience. The audience is arranged on three sides of the stage, and the fourth side usually includes a scenic background. In the proscenium arrangement the audience faces the stage from one direction and a curtain or *proscenium arch* hides all or part of the stage and the stage mechanism from the audience.

Theatrical convention, which means normal theatrical usage, calls for stage descriptions to be written for a proscenium stage. In a proscenium configuration the part of the stage closest to the audience is called *downstage,* while the part farthest from the audience is *upstage.* Those terms are holdovers from a time when stages were typically sloped, or *raked,* so the audience could see better. In most theaters now the audience area is sloped and the stage is flat, but the conventional terminology remains.

The side-to-side stage nomenclature is defined in terms of the performers rather than the viewers. Therefore *stage right* is actually the viewer's left and vice-versa. Your description of the setting should reflect those stage conventions.

The stage setting illustrated shows a room with a fireplace on the right wall. A door is up right. There are windows on the upstage wall, a bookcase up left and a second door downstage left. The drawing shows a couch at center stage with a chair to the left of the couch. Remember you only need to specify what is important. It may be enough to say simply: "The living room of a modern house." Edward Albee identified the setting of his *Who's Afraid of Virginia Woolf?* simply as: "The living room of a home on the campus of a small New England college."

As you work on the exercise in this chapter, you will find that characters begin to emerge in greater detail as they react within a specific environment, speaking to and struggling with other characters. That's a natural development, and the next chapter will deal more specifically with characters and various ways in which characters evolve.

In this same exercise you will be writing dialogue for the first time. A later chapter will go into more detail about dialogue, but for now it is more important for you to visualize and activate a conflict. Let what the characters say derive from what they want and from the kind of people they are.

At some point you may consider writing a screenplay instead of a play. There are many similarities in the two forms. Both plays and screenplays dramatize action. They both use conflict, character, and dialogue in similar ways.

There are, however, important differences. Although plays can certainly jump from place to place, they usually concentrate on one place and one set of characters for longer periods of time than do screenplays. I believe that a great part of the process of dramatic writing is learning to work with characters within particular situations, and I think you learn more about such fundamentals as character, dialogue, action, and conflict by working within the confines of a stage format because it forces you to explore relationships in greater depth.

Screenplays also rely far more than plays on the juxtaposition of visual images, particularly close-up visual images. A camera can show you a dripping faucet, followed by a gun lying on the floor, followed by drops of blood, and finally a corpse. Then it can cut to a man picking his teeth in a car. That kind of flowing imagery makes screenplays more like short stories or novels than plays.

Plays and screenplays employ very different writing formats, but that problem is easily overcome by looking at a book about screenwriting. If at some point you wish to tackle screenplays, I've listed several instructive books in the bibliography, but for the exercises in this book I would encourage you to stick with a play format. Except for a few instances where I have used a condensed format for a few lines, all examples in this book employ accepted playscript format, and you should get into the habit of using it, too.

And now let's proceed to the second exercise.

● ● ● **EXERCISE 2.1**
Place two characters in a scene of direct conflict. Write the scene in stage terms indicating how the scene would be staged. Use dialogue as needed, and resolve the scene. Minimum length: 2–3 pages.

The following scenes are examples of two-person conflict scenes written by students in a playwriting class.

EXAMPLE
              WHO'S SELLING OUT NOW?
                By Dave Dalton

         THE SETTING is a college dorm room. Rock
         and roll posters decorate the walls. A gui-
         tar and amp stand in one corner. On the

> bunk beds sit TIM and BRAD, college students about 18 years old.

                    BRAD

Listen man, the Grateful Dead couldn't play their way out of a paper bag next to Jimi.

                    TIM

Get real, man! Have you even been to a Dead show? Brad, they kick ass! Now the Who, they couldn't play their way out of a wet paper bag that was ripped all over.

                    BRAD

You have got to be kidding me! You're talking about the Who. The band that pioneered the rock opera. They were the first band to ever destroy their instruments on stage.

                    TIM

What are you talking about? They may have smashed a guitar or two, but I don't think they were the first band in the history of time to ever destroy their instruments.

                    BRAD

Yeah, well maybe they weren't the first, but they were the best. They turned smashing instruments into a friggin' art form. I mean, for real, Picasso was there in the front row of every Who concert taking notes on artistic expression.

                    TIM

Oh, get off it man! The Who is a bunch of old fusspots who sold out a long damn time ago.

                    BRAD

Sold out? The Who was one of the pioneers of rock, buddy. There is no way that they sold out.

                    TIM

Oh yeah? What about Tommy? What's that crap about doing it on Broadway with an orchestra?

                    BRAD

That's art, man!

                    TIM

It's a sellout. They wrote that thing in the sixties and turned it into some weird movie with Elton John in

big shoes. The point of rock and roll is rebellion,
man. It's rebelling against things like Broadway musicals. What could be more of a sellout than to take
something that you did in the sixties to rebel and turn
it into a money-grubbing Broadway hit? They're sell-
outs, man.

                    BRAD
Timmy, Timmy, Timmy, you're talking out your ass. In
the middle of their set at Woodstock, Townsend clubbed
a guy on the head with his guitar! Does that sound
like some kind of pacifist, sellout, money grubber? Then
at the end of the show he took the same guitar, which
he probably could have sold for thousands of dollars,
and destroyed it. You call that a sellout?

                    TIM
I'm just saying you never saw the Grateful Dead on
Broadway. You never saw Dylan playing his music with
some stuffy "wouldn't know rock and roll if it bit 'em
in the ass" orchestra. Those are artists, man. The
Who's just some spastic rock band with an extra large
budget for instrument replacement.

                    BRAD
Alright, man. I'm gonna show you how cool the Who are.

          (BRAD straps on his electric guitar and
          plugs it into his amp.)

                    TIM
Oh, please. I don't need some personal rendition of "My
Generation." Look, I'll admit the Who were rockers, but
they were never artists. They S-O-L-D O-U-T.

                    BRAD
No, man. I'm gonna show you art. Go open the window.

                    TIM
What the hell for? It's freezing out there.

                    BRAD
Go open the window. I'm gonna play a chord, and we're
gonna take this amp and throw it out the window.

                    TIM
What's that gonna prove?

                    BRAD
I'm gonna hit this chord, and then while it's still
resonating we'll throw the amp out the window. We'll
get the once-in-a-lifetime experience of hearing what

the Who heard. Townsend did this from a hotel room once. He said it was the most mind-altering experience he ever had. He said listening to that chord while the amp was falling and then hearing it explode in a rock-and-roll inferno as it hit the ground was better than, like, a thousand acid trips at once.

                         TIM
Man, are you sure you wanna smash your amp? I mean, what are you gonna play your guitar on?

                         BRAD
I'll get a new amp. This is a once-in-a-lifetime kind of thing, man. Plus it'll prove to your ass just what kind of artists the Who were. Now, open the window.

                         TIM
Alright, man, if you say so. Let's do it!

               (TIM opens the window. BRAD checks his gui-
               tar, fiddles with the amp, turning up the
               volume, and looks at TIM. They both nod.
               BRAD hits a chord; they both grab the amp
               and shout as they throw it out the window.)

                    BRAD AND TIM
The Who! The Who!

               (We hear the Doppler effect as the amp
               falls and then a small thump. BRAD and TIM
               stand looking out the window.)

                         TIM
Well, was it worth it?

                         BRAD

          (Shrugs.)

It sounded okay.

                    THE END

  EXAMPLE
                    ABSOLUTELY NOTHING
                      By Dana Cavallo

          THE SETTING is the living room of MEG's
          small apartment with couch and arm chair.
          The coffee table is littered with issues
          of Glamour, Vogue, and Cosmopolitan. MEG,

about 20, is sprawled on the couch blankly staring at the wall. JASON, also about 20, is in the chair reading the sports page of the newspaper. MEG fidgets. Eventually JASON looks up and glances at MEG.

                    MEG
What?

                    JASON
Nothing.

                    MEG
No, what?

                    JASON
Nothing, I said.

                    MEG
Well, then, why did you give me that look?

                    JASON
What look?

                    MEG
You know exactly what look I'm talking about.

                    JASON
Actually, Meg, I have no idea what you're talking about.

                    MEG
You gave me a look, Jason, and I just want to know what your deal is.

                    JASON
You want to know what my deal is. I'd like to know what *your* deal is! I was sitting here, just reading the sports page, and I looked over at you. No look in particular, Meg. It was just a "Hey, I wonder how Meg's doing over there on the couch" kind of look, and you freak out on me. So maybe I should be asking you what your deal is.

                    MEG
But something is bothering you, isn't it?

                    JASON
Yeah, Meg. You are bothering me.

                    MEG
I know what this is all about. You're angry about last night.

                              JASON
What about last night?

                              MEG
About me talking to that guy at that party.

                              JASON
What guy?

                              MEG
Oh, don't pretend like you didn't notice. Just that
guy. Just my friend.

                              JASON
Oh. "Just your friend," huh? So who is this friend?

                              MEG
His name is Bill.

                              JASON
Bill Anderson? Oh yeah. Good old Bill. Yep, he sure is
the friend type. That's fabulous. Bill Anderson. Just
great.

                              MEG
Jason, he's my friend. And we were just talking. And
you don't need to get jealous.

                              JASON
Uh-huh. And you were nowhere to be found for forty-
five minutes while I sat there with all your friends
trying to entertain myself. But, hey, you were just
talking to your old buddy, Bill, so what's the harm,
right?

                              MEG
Right. Just like there was no harm in you throwing
yourself all over Jessica Howlett.

                              JASON
Throwing myself at Jessica Howlett?

                              MEG
Well, you were with her most of the night, hanging all
over her.

                              JASON
The only reason I was even talking to Jessica Howlett
is because she was practically the only person I knew
there besides you. And you were making your rounds,
talking to everybody but me, hanging out with your
buddy, Bill Anderson, so what was I supposed to do?
Stand there like a loser?

                            MEG
You don't have a thing for Jessica?

                           JASON
Are you kidding me? Is that what this is all about?
She's one of the ditziest girls I've ever had the dis-
pleasure of talking to for forty-five minutes. I'm sure
she's not nearly as intriguing as Bill Anderson.

                            MEG
I was only talking to him to make you jealous.

                           JASON
Well, you succeeded.

                            MEG
Oh. Well, good.

              (They are silent for an awkward moment. MEG
              picks up an issue of Cosmo. JASON, con-
              fused, puts down his paper, fidgets, and
              stares blankly at the wall.)

                      THE END

EVALUATION

There are several instructive questions you might ask about these two scenes. First, do we know what the conflict is? Can we tell what each character's objective is? In other words, are the objectives clear and in opposition to each other? How is the conflict resolved?

Second, what do we discover about the characters through this conflict? Do we care about the characters? Do you want one character or the other to win?

Third, is there action in the scene? Can you state in a phrase what we see the characters doing?

Fourth, is the setting significant to the scene?

And finally, does the dialogue sound like people actually talking? Is the language appropriate for the characters in the scene? Do the characters talk differently?

On its surface, the first scene is merely an argument between two college boys about rock bands, but in an odd way it evolves into something more than that. The setting is an innocuous dorm room, and both boys appear comfortable there. Obviously the props of the amp and guitar play a prominent role.

The scene seems purely a verbal argument until Brad tries to prove his point to Tim, using the guitar and amp as his evidence. That action significantly raises the stakes, and at that moment the scene gains dramatic interest and begins to involve us in these boys and their argument. Will he really throw the amp out the window? What will happen if he does? Will that prove a point to Tim? Brad's decision to toss the amp even affects Tim, who mildly tries to talk him out of it. The climax to the scene, of course, is the amp tossing. While it doesn't prove which rock band is better, it does demonstrate the depth of Brad's commitment. It becomes his own symbolic rebellion against material possessions and his own vain attempt at an artistic statement of sorts. The resolution—that is, the boys' reaction to the event—seems rather anticlimactic and unsatisfying, as though neither the author nor the characters knew quite how to bring this startling event back into context.

Most of the boys' conversation displays a realistic edge. Notice, however, how the dialogue, when they are just verbally sparring, tends to come out in rather large chunks. But when Brad starts to do something, the dialogue exhibits a faster, give-and-take quality.

The place has little impact on the gentle second scene, which features very little overt action, but a large psychological movement. Although on one level "absolutely nothing" happens, Meg has a bone to pick, and she initiates the conflict. In fact, both characters are upset about the actions of the other at a party the previous night. Both want reassurance that they are loved, one of the most basic of human desires, and it is that common need that interests us in the characters. Their language, which flows nicely back and forth in a youthful manner, skirts the issue until Meg confronts it. Interestingly, rather than address her own concerns directly, she opens the door by insinuating a problem for Jason: "You're angry about last night." Finally, Jason directly accuses Meg of leaving him stranded, and Meg retaliates by expressing her jealousy over Jessica. Eventually, Meg gets her reassurance, but Jason is left wondering about Meg's intentions. The final tableau reveals a nice reversal of the opening as Meg reads contentedly while Jason fidgets and worries.

As they created conflict, both writers also began to generate characters. Now that you've seen how any two characters can come into direct conflict, let's take a more detailed look at exactly how characters develop.

3

• • • • • • • •

Character

EXERCISE 3.1 • CHARACTER
List the characteristics of three people you know very well.
Select one of those people, and place a character based on that
person in a situation of conflict with another person. Use
correct stage terminology and dialogue as needed. Minimum
length: 2–3 pages.

In the previous chapter on conflict, I stated that conflict does
not arise out of nothing; rather, it stems from intentions at cross purposes. Intention, in turn, stems from character. A logical question follows:
If conflict comes from intention, and intention comes from character,
why start with conflict? Why not start with character? Many writers, as
you can see from the Character vs. Action chart (see following page),
ask just that question and argue for the preeminence of character. Lajos
Egri, a significant and sensible writer on playwriting, concluded, "There
is no doubt that conflict grows out of character." Why not, then, begin
there? The answer is this: Character does not necessarily produce
obstacles or conflict, but obstacles and conflict necessarily bring out
character.

Let's assume you've thought of two characters for the conflict
exercise—a doctor and a trash collector, for instance. What do we know
about them? Nothing. Now let's put them in a situation of conflict. Using
the prompts suggested, we can first envision a place where the two characters come together, and we'll use the occupation of the trash collector. He's
making his rounds, and he comes to the house of the doctor.

Now imagine an object that could force their conflict. The trash
collector is picking up bags, and a plastic bag bursts, littering debris over
the driveway in front of the doctor's expensive house. The trash collector
sighs and mutters an expletive under his breath. Suddenly a man emerges
from the house and hurries to the trash collector.

Doctor: Hey, what are you doing here?

Trash collector: The bag broke.

CHARACTER VS. ACTION

For Character	*For Action*
The dramatist who hangs his characters to his plot, instead of his plot to his characters, ought himself to be hanged. <div align="right">*John Galsworthy (paraphrased)*</div>	Everything hangs on the story; it is the heart of the theatrical performance. For it is what happens *between* people that provides them with the material to discuss, criticize, alter. <div align="right">*Bertolt Brecht*</div>
My plays deal with people, and thinking, and believing and philosophizing are all, to some extent at least, a part of human behavior. <div align="right">*Friedrich Duerrenmatt*</div>	The dramatist must be by instinct a storyteller. <div align="right">*Thornton Wilder*</div>
Before I write down one word, I have to have the character in mind through and through. I must penetrate into the last wrinkle of his soul. I always proceed from the individual; the stage setting, the dramatic ensemble, all of that comes naturally and does not cause me any worry, as soon as I am certain of the individual in every aspect of his humanity. <div align="right">*Henrik Ibsen*</div>	Things occur to me first as scenes with action and dialogue, as moments developing out of their own vitality. <div align="right">*George Bernard Shaw*</div> Plays should deal with moments of crisis. <div align="right">*Marsha Norman*</div>
The difference between a live play and a dead one is that in the former the characters control the plot while in the latter the plot controls the characters. <div align="right">*William Archer*</div>	A play lives by suspense, and suspense comes from complication. <div align="right">*Kenneth MacGowan*</div>
Dream out a story about the sort of persons you know the most about and tell it as simply as you can. <div align="right">*attributed to Owen Davis*</div>	History shows indisputably that the drama in its beginnings, no matter where we look, depended most on action. <div align="right">*George Pierce Baker*</div>
Once I know what my characters are doing, the play comes very easily. <div align="right">*Terrence McNally*</div>	Wherever you start, eventually the material must take on some sort of shape. In order to give it shape, you have to get some type of story. <div align="right">*Josephine Nigli*</div>
Every great literary work grew from character . . . character creates plot, not vice versa. <div align="right">*Lajos Egri*</div>	
I'm not following a plot line so much as I'm following the surrender of my audience's	Drama is about what happens

emotions to the dynamic of the realities of my characters.

Ntozake Shange

First—and this is terribly important—we get to know our characters. We try to get ourselves out of the way and let our characters live.

Jerome Lawrence

next, and if I don't know what's going to happen next . . . then I don't think the play will have the necessary momentum.

A. R. Gurney

Structure is storytelling. You are building toward something at the end of the play. That's why I don't write until I know what I'm going to do at the end of the play.

Robert Anderson

I like characters that help the plot along and keep it moving and let us know where we are.

John Guare

Doctor: I can see that. Are you just going to leave this garbage all over my front yard?

Trash collector: Look, Mac, you got a million-dollar house there, so how come you use cheap bags? It ain't my fault it broke.

Doctor: I got it out here, didn't I? It didn't break on me.

Trash collector: I don't need this, buddy.

Doctor: You get it cleaned up. Every bit of it. I've got to take out a gall bladder in twenty minutes, and I can't even get my car out of the driveway.

Trash collector: So call a cab.

Doctor: You get it picked up or I'll report you!

Trash collector:
> (Throws down the remnants of the bag in his hand and signals for the truck to pull on.)
Pick it up yourself, Mac. And next time use a better bag!

It may not be scintillating theater, but in the course of this small conflict, two characters—an arrogant doctor and a proud, defensive trash collector—begin to emerge.

As this example illustrates, you cannot create a dramatic character in isolation. You may list a multitude of characteristics for your characters:

age, occupation, physical appearance, favorite activity, recently read books, likes and dislikes, and so on. But the characters won't really begin to reveal themselves until you place them in dramatic situations rife with action and conflict.

The revelation of character through conflict and action in drama should hardly come as a surprise, for the process is identical to what occurs in real life. You don't simply meet a person and instantaneously know that person's character. Rather, the character of individuals emerges bit by bit through their actions, through what they do and say, and through their interactions with other people. Character emerges in drama in just the same way.

Let us for a moment consider *you* as a character. Imagine that you're driving down a winding two-lane highway. A bystander might observe aspects of your personality. Are you driving a sparkling new BMW or a rusted pickup truck? Are there any bumper or window stickers espousing causes, identifying a school, or indicating a special parking permit? Are you wearing a tank top or more dressy attire? Do you drive cautiously on the turns, or do you attack them like a race-car driver? All of those elements provide clues about you.

Now let's complicate your life. Suppose the road is icy. Do you sit up in the driver's seat? Do we read concern on your face? Perhaps you relish the challenge. What if another car cuts you off? Do you curse or make a gesture at the other driver? Or do you just take a breath and collect yourself? Imagine that your car breaks down or has a flat tire. Do you know what to do to fix the problem yourself? Do you hail a passing motorist? In these instances the conflict generated by the obstacles of weather, another motorist, or the inanimate automobile itself, and your responses to those challenges, would reveal additional facets of your personality.

As you saw from the Character vs. Action chart, critics have expended an enormous quantity of ink in a long-running debate over whether character or action is more important to a play. No one has expressed the case for the dominance of characters over action more eloquently than John Galsworthy, an early 20th-century playwright. He wrote:

> In drama, undoubtedly the strongest immediate appeal to the
> general public is action. . . . The permanent value of a play,
> however, rests on its characterization. Characterization focuses
> attention. It is the chief means of creating in an audience sympa-
> thy for the subject or the people of the play.

More succinctly, Galsworthy said, "A human being is the best plot there is." That point of view has many supporters, including Henrik Ibsen, Ntozake Shange, Friedrich Duerrenmatt, Terrence McNally, and Jerome Lawrence.

But if writing plays were a debate among authorities, the case for the preeminence of action or plot could be argued by Bertolt Brecht, Thornton Wilder, Robert Anderson, A. R. Gurney, and others. Wilder, for instance, concluded: "Drama on the stage is inseparable from forward movement, from action."

Unfortunately, as is the case with other long-standing questions such as "Which came first, the chicken or the egg?" the question "Which is more important, character or action?" is, finally, an empty one. Character and action are inseparable. It is as impossible as it is undesirable to have one without the other.

A scene or a play can begin with a story or it can begin with an interesting character. Moreover, a play can begin in many other ways as well. It can start with a word or a phrase. It can be propelled by a firmly held conviction. Tina Howe once said that she always started a play with a setting. Her play *Painting Churches* takes place in a Boston interior while her *Coastal Disturbances* takes place on a beach.

Wendy Wasserstein has said that her plays often spring from a visual image. Her Pulitzer Prize–winning piece *The Heidi Chronicles* began with the picture of a woman standing in front of a group of other women and saying that she'd never been so unhappy in her life. Jean-Claude van Itallie echoes Wasserstein and Ingmar Bergman when he states that, for him, the seed of a play is a provocative and mysterious visual image "which I can turn like a prism or a crystal in the light. . . . The questionings of that image are the beginnings of the play."

What's important is not where plays begin but where they end. No matter what they start with, somewhere along the line of their development they must incorporate all the elements of an effective play—well-defined characters speaking interesting dialogue in a forward-moving story, all brought together within an enticing environment.

To demonstrate how different elements influence each other, particularly how action and character reinforce one another, let me describe an action. This action was written by a student in a playwriting class.

EXAMPLE

A WOMAN AT CHRISTMAS
By Mary Parker

A middle-aged woman stands scrutinizing a Christmas tree, which stands undecorated in the corner of a small room. Deciding that it is leaning toward the left a fraction, she walks determinedly to it, kneels, and presses the trunk toward the right while she tightens the screws of the stand on the left side. She stands, takes a few steps backward, runs her fingers through her graying hair, and observes the tree for a few seconds. Sighing, she walks slowly from the room and returns with a stack of boxes. As she stoops to place the boxes down in a chair beside the tree, the top box falls off, and shattering glass is heard as it hits the floor. She sets the other boxes down and then bends slowly over the fallen box. She breathes in deeply and removes the top of the box. Upon seeing the contents, she sits on the floor and exhales.

Hesitantly, she reaches into the box and brings forth a large broken piece of shiny green glass. She holds it up toward the light and watches the light sparkle off the broken edges. She places it on the floor by the box and gets up. She walks out of the room without looking back and slowly shuts the door behind her.

THE END

That action not only exudes a somber tone, it conveys a sense of character. The middle-aged woman is alone at the holidays. Her unsmiling expression and regular pace alert us to her depression. Her concern that the tree be properly upright informs us that this is a precise, meticulous woman. The scene suggests that she is attempting to keep a hold on her lonely life through the ritual of trimming the tree. But when the ornament box falls and a special ornament is broken, she becomes desolate and cannot sustain the pretense. She gives up and leaves the room.

Now let's turn around and describe a character. I'm thinking of a boy of about ten. He has a constant runny nose, which he wipes with the back of his hand. This young man wants to be thought of as tough. Even on the coldest days he leaves his shabby coat unzipped. He is something of a bully, taunting the younger children at school and throwing snowballs at them.

Here we have the seed of a character, and even in this early stage the character begins to emerge in terms of action and conflict: what he wears and how he wears it; his personal behaviors; what he does to other people.

You probably discovered as you wrote your action and direct conflict scenes that they, like the paragraph about the woman at the Christmas tree, began to suggest moods and convey character. That occurs not because action and conflict are more important than other elements, but because the various elements of drama work together, support each other, and reveal each other.

That is what happens in the action of the woman trimming the tree. Her determined walk and unsmiling visage work together with her gray hair to suggest a certain character. That, in turn, is supported by her slow, hesitant movement and careful breathing in response to the fallen ornament box. And all of those elements, juxtaposed within what should be a joyous event, create a revealing, dramatic moment.

Although you cannot create character in isolation from other elements, you can use characters as the impetus for a scene or a play, and one way to do that is to write about characters with whom you are familiar.

• • • **EXERCISE 3.1**

List the characteristics of three people you know very well. Select one of those people and place a character based on that person in a situation of conflict with another person. Use correct stage terminology and dialogue as needed. Minimum length: 2–3 pages.

Begin your lists with physical descriptions. Then list distinctive vocal characteristics or particular modes of speaking. Add to your list other external information such as age, occupation, religious affiliation, race, and origin. List particular likes and dislikes of the person. Then add psychological factors that reflect the person's values. What bothers this person? What does he or she care the most passionately about? What would upset this person?

When you have completed three lists, select *one* of these people to develop into a character, and you're ready to start your scene.

EXAMPLE

BIG BROTHER

By Jane Rupp

THE SETTING is a kitchen. SARAH, about 15, studies at the table. RICK, quite a bit larger and a year or two older than SARAH, ENTERS with the comic section of the newspaper. He sits facing SARAH, looks at the comics for a moment, then addresses SARAH.

RICK

Hey, Sarah, whacha doin'?

SARAH

Well, Rick, what does it look like I'm doing?

RICK

I dunno. Looks like homework, right?

SARAH

Yep. Definitely doing homework.

RICK

Oh. Did you get your hair cut or something? It looks different. Nice, I mean.

SARAH

(Closes her book and lays down her pen.)

Okay, Rick. What do you want?

                              RICK
What are you talking about, Sarah?

                             SARAH
You know what I'm talking about. Now what's up?

                              RICK
Well, I guess I was just wondering—if you're not too
busy, that is—if you could make me lunch.

                             SARAH
And what, may I ask, is wrong with your legs and arms
that I have to make you lunch? You're not even doing
anything worthwhile!

                              RICK
Oh, and like you are. Please.

                             SARAH
                  (Returns to her homework.)
Nope. Sorry.

                              RICK
C'mon, Sarah. Why not. I'm nice to you. I <u>deserve</u> it.
                  (SARAH snorts in response to that comment.)
Pleeese. Just this once, Sarah. I'll repay you, I
promise.

                             SARAH
                  (Slams book shut and glares at RICK.)
Let's just get one thing straight. You will <u>never</u> repay
me, and there is absolutely nothing that I can think of
that you have ever done to deserve my kindness. You are
only nice to me when you want something from me. Now,
get off your lazy butt, and make your own damn
sandwich.
                  (SARAH returns to her book, furiously turn-
                  ing pages.)

                              RICK
You don't really mean that. C'mon, I'll help you with
your homework.

                             SARAH
How could you possibly help me with my homework?

                              RICK
Well, at least I offered. See? Now we're even. You
practically <u>owe</u> me lunch!

                             SARAH
                  (Glances at the clock on the wall.)

Will you leave me alone if I make you your stupid lunch?

                    RICK
But of course, my dear, kind sister.

                    SARAH
And will you let me ride to school with you tomorrow morning?

                    RICK
Absolutely. Only morons take the bus.

                    SARAH
          (Gets up and goes to the refrigerator.)
Ham and cheese okay?

                    RICK
Whatever you make is fine with me. I'm easy to please.

                    SARAH
          (Mumbling to herself.)
Why do I do this?

                    RICK
What did you say?

                    SARAH
Oh, nothing. I just wondered if you realized that loaves of bread don't come with the meat already in them.

                    RICK
Oh, you're a stitch. Boy, you always make everybody pay for the slightest little favor. But it really doesn't matter. I'm sitting down reading the comics, and you're making me food. Life is good, and you can say whatever you like.

                    SARAH
Ooooh, Rick, you are so clever.
          (SARAH sets a sandwich in front of RICK.)
Now remember, you're taking me to school tomorrow.

                    RICK
          (Takes the sandwich and heads toward the
          door.)
Sheeyeah, right! Ride the bus, moron!

                    SARAH
See, I told you you'd do that!
          (Sits at the table.)
I am such an idiot to listen to him.

                          RICK
          (Offstage)
Yeah, you're right there!

                        THE END

EXAMPLE

                    CONVERSATION
                   By Amy Feezor

          (THE SETTING is a small kitchen. Center is
          a rectangular table with two matching
          chairs. NATHAN sits at one of the chairs,
          writing on a pad of paper. ELLEN walks in
          through the main door, stage right. She's
          holding a grocery bag, and a black purse
          hangs from her arm. As she sets down the
          bag and purse on a counter, ELLEN addresses
          NATHAN.)

                         ELLEN
Nathan.

                         NATHAN
          (Continues to write.)
Where've you been?

                         ELLEN
          (Putting groceries away.)
Grocery store on George Street. They were having a sale
on broccoli, and you know how I'm on that broccoli
kick, so I went to the store down there.

                         NATHAN
Uh-huh.

                         ELLEN
It's a pretty nice little place. I might go there more
often. They have the sweetest little grey-haired man
who runs it.

                         NATHAN
Really.

                         ELLEN
          (Bringing a glass bowl and a plastic bag of
          candy to the table.)
I think he has a thing for me.
          (She puts the glass bowl in the middle of
          the table. NATHAN looks up.)
You know, like a crush? He called me "darlin'."

NATHAN
(Resumes writing.)
Who?

ELLEN
The sweet little gray-haired man. I was his darlin'.
(Rips open the bag and pours pastel-
colored, heart-shaped candies into the glass
bowl. NATHAN looks up and watches her.)

NATHAN
What <u>are</u> you doing, Ellen?

ELLEN
(Finishes pouring, wads up the pink plastic
bag, and throws it away.)
I bought us some conversation hearts. You know, those
little Valentine-hearty candy things. They were on
sale.

NATHAN
What are we going to do with them?

ELLEN
Hello. Nathan. We're going to <u>eat</u> them, silly.
(Starts to pick through them.)
Have some.
(NATHAN grabs a handful and starts to pop
them in his mouth.)
WAIT!
(He stops.)
You don't shovel them in your mouth like they're Chex
mix. They're <u>conversation</u> hearts. You have to read them
first.

NATHAN
Ellen, there's got to be five pounds of candy in that
bowl. It would take us fifty years to read them and
then eat them all.

ELLEN
(Ignoring him, picks a candy from the
bowl.)
Look. Look. This one says HOT STUFF. Hey, that's you!
(She smiles at him. NATHAN smirks, pours
the candies from his hand back into the
bowl and resumes writing. ELLEN eats the
candy and picks another.)
Ooo! And—oh, how cute. This one is MY SWEET.
(NATHAN writes. ELLEN chews and watches
him. She picks another candy.)

Oh dear. Uh-oh. Nathan, lookie here. Look here!
>   (Tries to show him the candy. He still
>   writes.)

This one says MY DARLING. Oooo my, just like that sweet
little old man at the store. "Goodbye, Darlin'" he said
to me. Goodbye, Darlin'. Now you can't tell me that
you're not just a <u>little</u> bit jealous, right, Nathan?
>   (She reaches over to pinch his cheek, but
>   he pushes her arm away.)

                              NATHAN
Ellen! Cut it out!

>   (He glares at her, then resumes writing.
>   She stares at him. He writes. Slowly she
>   reaches to the bowl, takes a handful of
>   hearts and begins to eat them without look-
>   ing at them.)

                              THE END

EVALUATION

In the first example, the student used her own "Big Brother" as the germ for the scene. Rick is certainly dominating and irritating, but perhaps the most intriguing part of the scene is the way in which Sarah allows herself to be manipulated. She *knows* Rick will abuse her good nature, but she goes ahead anyway, almost as though she feels a compulsion to do things for him even though she knows he'll exploit her. She seems to *want* to be a martyr. As for Rick, he enjoys the contest. Clearly it would be simpler for him to make the sandwich than to get his sister to do it for him—but there's no challenge to that!

The structure of the scene works well. Rick interrupts Sarah and badgers her. She finally agrees to his entreaty, but only *after* she achieves what she thinks is a bargain. Then the playwright crafts a *reversal,* where the scene seems to go in one direction, then abruptly changes course. It appears Sarah and Rick have reached a compromise, but after Sarah does her part, Rick reneges and gloats in the fact that he has taken advantage of his sister yet again. Dealing as it does with sibling rivalry and with one person trying to take advantage of another, the scene probes relationships with which we can easily identify.

The second scene, another example of what is sometimes called "kitchen table" drama because it treats basic domestic relationships, could take place in virtually the same kitchen as *Big Brother.* The scene sprang from an observed relationship, and its charm results directly from the use of the prop—the little candy hearts—as conversation starters. One character is trying desperately to create a conversation. The other is just as intent to avoid it. We don't know what Nathan is writing, but we can see clearly that Ellen is not a part of his activity. The turning point comes when Nathan takes some candies but refuses to engage in the conversation that Ellen insists goes with them. Here, too, we see a reversal. It seems the two will come together over the hearts, but instead Nathan simply drops his candies back in the bowl, and the defeated Ellen winds up eating the hearts without even reading them.

Significantly, both scenes utilize props and basic but important actions. Sarah's making the sandwich is a central development, and the candy hearts are especially well chosen because they're something everyone knows and they resonate with notions of Valentine's Day and metaphors of love.

Playwrights often use props or games to provide dramatic tension, to act as a metaphor, or to help provide a structure. The central focus of *The Gin Game*, the Pulitzer Prize–winning play by D. L. Coburn recently revived on Broadway, is an on-again, off-again card game played by an elderly man and woman. In Tina Howe's *Road to Zanzibar* the characters play a game called Geography, which leads to the title of the play. In *A Streetcar Named Desire*, Stanley's poker night serves as an emotional backdrop for the climactic moment of the play.

The dialogue of both scenes, while basic, flows naturally. Ellen's recounting of the old man calling her "Darlin'" and the MY DARLING candy at the end provide clever and meaningful brackets to the scene.

In both scenes characters are revealed because of what they want and how they go about getting it. Not only does Rick want the sandwich, but he wants to trick his sister. And he succeeds. Ellen wants to start a conversation and revive a relationship, but she fails, and the failure seems to take the heart out of her.

Now that we've created some characters to inhabit the worlds we invent, in the next chapter we'll work on how to provide characters with effective voices—in other words, how to write dialogue.

4

• • • • • • • •

Dialogue

EXERCISE 4.1 • DIALOGUE

Listen to and write down the dialogue of two selected individuals. Write a scene of conflict in which you incorporate into your dialogue some of the speech habits you've observed. Minimum length: 4–5 pages.

Just as character stems from action and conflict, so, too, dialogue emerges from conflict and action and character. It does not just exist by itself. If you try simply to create characters talking, they will have little to talk about, and the resulting dialogue will veer toward pretentious philosophizing or empty small talk. If, on the other hand, you create the dialogue to serve a purpose, to complement an action, or to gain a victory in a conflict, it will be necessary and important.

In some ways dialogue is a trap. Because we are so accustomed to thinking of plays as dialogue, we run the danger as writers of concentrating only on the words and forgetting the significance of the action, the conflict, and the characters. Good dialogue should be a means for a character to accomplish an end, so don't lose sight of the goals that your characters mean to achieve.

Some playwrights have particular difficulty writing dialogue. Instead of writing as people speak, they write as if they were penning an academic composition: complete sentences, proper grammar, no contractions. Playwriting students often seem to write dialogue more formally than they write term papers.

There are, of course, some basic guidelines for dialogue that can assist beginning writers. Konstantin Stanislavsky, the famous Russian director and the man credited with developing an approach to realistic acting, once said that all he wanted was for performers on stage to act like people in real life. That sounded simple, but it turned out to be more difficult to achieve than it sounded.

Writing dialogue is much the same. All we want for realistic dialogue is that characters in plays talk the way people talk in real life. That means they use contractions, incomplete sentences, repetitions, fragments of thought, perhaps bad grammar, and all the other imprecisions of expression to which we humans are subject. The doctor and the trash collector in their brief encounter in the previous chapter used contractions ("I got it out here, didn't I?"), fragments ("Every bit of it") and bad grammar ("It ain't my fault").

Larry Shue, the comic playwright of *The Nerd* and *The Foreigner*, observed how people begin to say something one way, then reverse themselves and formulate the words another way around instead. In *The Nerd*, Rick Steadman has just made a mistake in a game when Tansy says to him: "You can't—maybe I didn't explain this; see, you're supposed to say something beginning with an *E*." In another place a character asks: "What're we—? What's the point of this?"

In a more serious vein, Anna Deavere Smith initiates one character's monologue in *Twilight: Los Angeles, 1992* with:

> Our life is something like,
> uh,
> what's the name of that picture
> with Dorothy Dandridge
> when she was like a prostitute and the guy she met was in
> the Air Force—the service?

With their stops and starts, both Shue and Smith capture the way people actually put thoughts together.

Different characters should talk in different ways, just as the doctor and the trash collector did. "I don't need this, buddy" is a line only the trash collector would say, not just because of the content but because of the familiar use of "buddy." And "Are you going to leave this trash all over my front yard?" could only belong to the doctor, not just because of the content but because of the formal construction of the sentence.

Of course, such things as level of education, intelligence, and environment affect expression. But even with two characters from the same area who are roughly equivalent in intelligence and education, there should be differences. One person may be assertive or optimistic ("Let's open a store!") while another is submissive or pessimistic ("Maybe we could open a store?"), their speech should reflect those qualities.

Remember, too, that people in real life don't generally talk in speeches. Just as effective acting demands genuine interaction between the performers, so dialogue should be a two-way street, with give and take between the characters. Frequently in good dialogue a word or phrase spoken by one character is picked up and repeated by another. Each statement acts as a stimulus causing a response that in turn becomes a stimulus causing a response and so on.

David Mamet is praised for writing realistic contemporary dialogue. In fact, he has been known to frequent restaurants and record the way people actually talk. The result is halting dialogue that moves around a subject, dialogue that allows performers to supply nuances to words through the way in which the lines are delivered. This opening exchange from Mamet's *American Buffalo* illustrates his technique. Don operates Don's

Resale Shop where the action is set. Bob is a young friend whom Don has sent to follow a certain man.

```
Don:   So?
               (Pause)
       So what, Bob?
               (Pause)

Bob:   I'm sorry, Donny.
               (Pause)

Don:   All right.

Bob:   I'm sorry, Donny.
               (Pause)

Don:   Yeah.

Bob:   Maybe he's still in there.

Don:   If you think that, Bob, how come you're here?

Bob:   I came in.
              (Pause)

Don:   You don't come in, Bob. You don't come in until
       you do a thing.

Bob:   He didn't come out.

Don:   What do I care, Bob, if he came out or not?
       You're s'posed to watch the guy, you watch him.
       Am I wrong?

Bob:   I just went to the back.

Don:   Why?
               (Pause)
       Why did you do that?

Bob:   'Cause he wasn't coming out the front.

Don:   Well, Bob, I'm sorry, but this isn't good enough.
       If you want to do business . . . if we got a busi-
       ness deal, it isn't good enough. I want you to
       remember this.

Bob:   I do.

Don:   Yeah, now . . . but later, what?*
```

Note how Bob's "there" is reversed by Don's "here," and how the verbs "come," "came," and "went" are introduced, picked up, and repeated over seven lines. Mamet's ear for the repetitions, juxtapositions,

*By David Mamet. Reprinted by special permission.

and the non sequiturs of human communication provide a natural, flowing sound

Some individuals display a better sense of words and the way people use them than others. That's to be expected, just as some people demonstrate a finer facility for devising plots than others. Eugene O'Neill in the early stages of his career gained fame for writing realistic dialogue that captured the accents and the flavor of the seamen who populated his first plays. Later in his career he was criticized for overwriting his dialogue and for lacking a poetic sense of language. Yet some of his later plays, such as *A Long Day's Journey Into Night*, are among his most powerful. Tennessee Williams's poetic use of words and rhythms within the context of action and character is one of his greatest assets. In any case, don't think you're automatically a playwright because you discover you have a facility for dialogue, and, conversely, don't despair of writing plays because your dialogue doesn't immediately sparkle.

EXPOSITION AND "CHUNKING"

One of the hardest tasks about writing dialogue is gauging how much information to include, for almost every play needs to inform the audience about certain facts. Background information about characters or plot is called *exposition*, and good playwrights try to include exposition as naturally as possible. Also, as the play progresses, the author must reveal various elements of the plot or story to the audience. There, too, the aim is to provide needed information as an unobtrusive part of the natural flow.

Playwrights often cram too much exposition into a character's line, and the result is wooden, information-laden speech. I call that problem "chunking" because information spills out in large chunks rather than in bits and pieces.

Look at this opening passage to a scene in a girl's bedroom. Melanie, a small woman about 40 who is dressed in black, walks slowly around the room looking at several items. Finally she picks up a stuffed bear from one of the twin beds just as Sandra, also a small woman in black, enters.

Sandra: Hey, what's wrong? How come you came upstairs?

Melanie: It's just so crowded down there, and most of those people didn't even know Daddy. They just came to watch us and see if we'd cry.

Sandra: You sound bitter.

Melanie: I shouldn't have come home. I always get crazy when I come back.

Sandra: Maybe if you'd stayed around long enough you'd have seen it wasn't so bad. You come home knowing everything will be bad, so it is. Once you got out you never came back just to <u>observe</u> the situation. And off you'd go again.

Melanie: Sandy, that's just not fair. You know it's hard for me to be here. I remember all the times Mom and Dad fought. Sometimes I can still hear them screaming. It never got any better, Sandy. I hate coming here. There have been times, though, when I've wondered if I should have stayed so you could have left.

Almost every line is packed with information. We really don't need to get everything all at once. In addition to the chunking, the speeches are quite long with no give and take, and the emotions are very directly confronted. Those problems give the scene a feeling of wooden characters spouting mechanical lines. Let's look at the same scene with the emotions approached somewhat less directly and the lines broken up and varied.

Sandra: I figured I'd find you here.

Melanie: I've always liked this room.

Sandra: You found Jelly Bean.

Melanie: I don't believe they kept him all these years.
　　　　(Pause)

Sandra: Crowded downstairs, huh?

Melanie: Most of them didn't even know Daddy.

Sandra: They're just trying to be nice. They want to make sure we're okay.

Melanie: They want to watch us grieve.

Sandra: Mellie.

Melanie: It's true. They just want to see tears.

Sandra: Well, they wouldn't see any from you.

Melanie: Meaning?

Sandra: Nothing.

Melanie: I cared.

Sandra: Did you?

Melanie: I shouldn't have come home.

Sandra: For the funeral? I don't believe you said that.

Melanie: I always get crazy when I come back.

Sandra: You always come back <u>knowing</u> everything will be
    awful, so it is.

Melanie: Sandy, that's not fair.

Sandra: Why can't you just <u>look</u> for a change instead
    of <u>judging</u> . . . .

Melanie: All I saw was Mom and Dad fighting. Daddy's
    dead, and I still hear them screaming at each
    other.

Sandra: It's over now.

Melanie: You shouldn't have to hate coming home.

Sandra: I know.

Melanie: (handing the bear to Sandy) Sometimes I wonder
    if maybe I should have stayed so you could have
    left.

That scene continued as the two sisters tried to reach common
ground, but I suspect the opening sequence is enough for you to see how
the student author parceled out information in small segments that lead the
audience on instead of showering them with everything at once.

How would you rework the chunked parts of the following
short passage? Amy and Cassie are sitting at a table in an ice cream
shop.

Amy: You know, Cassie, I get so depressed when I think
    of this summer. I don't know if I can last with-
    out Brian. Sometimes I wish I had never met him,
    and other times I don't know if I can live with-
    out him.

Cassie: Wait a minute, Amy. Weren't you just recently
    telling me that you had had it with Brian and
    that you were ready to see someone else?

There are several problems just in the first two lines. First, gobs of
information are chunked together. Also, the dialogue is extremely direct.
The character's problem arises immediately, and both characters express
exactly what is on their minds. Furthermore, they use long speeches with
relatively common, clichéd language. Next, the setting isn't used at all.
There's no apparent reason to be in an ice cream shop.

Finally, I'm uneasy whenever I hear lines like "Weren't you recently telling me . . ." Beware of dialogue you write that starts with "As I
said before" or "You know that" or "I already told you." Such lines usually
indicate that the characters already know what's being said, so it's actually
only being repeated for the benefit of the audience. Information that is

rehashed must contribute something new—a discovery of something in the old information or a new angle for a character. Otherwise it's just clumsy exposition.

How did you rewrite the two lines from the ice cream shop? They might be better if they went something like this:

```
Cassie: Good soda, huh?
          (No response)
      Hey, it's summertime. You're supposed to be
      happy.

Amy:  I was thinking about Brian.

Cassie: Brian? I thought you'd had it with Brian.

Amy:  Well, yeah. I mean, sometimes I wish I'd never
      met him, but other times I don't know if I can
      live without him.
```

There's a better flow to those lines, some sense that the characters are actually paying attention to each other, and there's at least minimal use made of the setting.

CLICHÉS

The phrase "I don't know if I can live without him" still smacks of cliché. A cliché is a word or phrase that has been so frequently used that while the meaning is clear, the words convey little about the character who speaks them. If you find clichés in your dialogue, examine what the character is actually saying. For example, if a character says "My hands are tied," the actual meaning is, "Someone above me won't let me do anything about it." Reduce the cliché to the basic meaning. Sometimes that basic language will provide the best means of expression. Other times it will guide you to find original words appropriate for a particular character to express the meaning.

Occasionally clichés can be effective if they fit the way a particular character would talk. There are reasons, after all, why clichés become clichés. They often express a complex thought in a few words. Still, good playwrights try to invest even their most pedestrian characters with distinctive means of expression. In Tina Howe's *Coastal Disturbances*, the lead character of Holly expresses a sentiment very similar to Cassie's when she says of a young man she's met: "He makes me crazy, but I'm just so alive with him." If your character must use a cliché, perhaps you can devise an interesting way to insert it. Amy and Cassie's dialogue might go like this:

```
Amy:  Well, yeah. I mean, sometimes I wish I'd never
      met him, but other times he's so . . .

Cassie: Divine?

Amy:  No. He's so . . .

Cassie: Heavenly?
```

Amy: No. Neal!

Cassie: Oh.

That makes Amy the queen of triteness.

Writers create natural dialogue and avoid clichés by being careful listeners and observers of the world around them. Just as David Mamet recorded conversations in restaurants, so you, too, should train your ear to hear the peculiarities of speech of the people around you. Then you should write down your observations. Virtually every good writer I know keeps a notebook with him or her at all times to jot down observations, turns of phrase, or ideas. Some writers have drawers full of their old notebooks, and if they hit a block, they turn to those for an idea to get them going again. If you want to be a writer—any kind of writer—you should cultivate the practice of keeping a notebook.

If you aren't already keeping a notebook or journal, perhaps the following exercise, which is designed to increase your sensitivity to the way people express themselves, will get you started.

• • • EXERCISE 4.1
Listen to and write down the dialogue of two selected individuals. Write a scene of conflict in which you incorporate into your dialogue some of the speech habits you've observed. Minimum length: 3–4 pages.

After you've selected your two individuals, listen carefully to the way they talk. Listen for and write down incomplete sentences and repeated words and phrases. Note particular rhythms and cadences, pauses, inflections, and unusual word choices. Look for nonverbal contributions to the conversation such as gestures, looks, and inflections. You'll be amazed at what you discover not only about speech but also about characters and relationships.

ADVERB DISEASE
Before you begin your scene, let me add just one more suggestion. Inexperienced writers frequently try to refine their dialogue by describing in stage directions how the line should be delivered. This "adverb disease" results in lines like:

GEORGE (forcefully): Don't go in there!

While I'm a great believer in action-oriented stage directions, such temperament descriptors are seldom necessary, and many performers and directors take them almost as an insult. If your dialogue and your character aren't sufficient, it's highly unlikely that adverbs such as "gleefully," "menacingly," or "brightly" will rescue them. Rather, take your cue from Terrence McNally, who once boasted that the only stage directions he ever writes are "Enters," "Leaves," or "Dies."

EXAMPLE

GOOD IDEA, IF IT WORKS
By Dwayne Yancey

THE SETTING is a gas station at night. At right is a large front window and the door, which leads out to the pumps. In addition to the weak fluorescent lights overhead, there is a pale light shining through the windows from the station sign outside. There is a pyramid of oil cans and a tire display in the front window. A large side window covers upstage. In the center is a Coke machine, with A SCRUFFY-LOOKING TEEN-AGED BOY leaning against it. He is dressed in torn and faded blue jeans, an old T-shirt, a denim jacket, and a toboggan-style cap. He holds a can of Coke in his hand, his arms crossed, and he is looking at the floor. At left is a display case filled with assorted candies; on top is a cash register and a stack of folded maps. Sitting on a stool behind the register is ANOTHER SCRUFFY-LOOKING TEEN-AGED BOY, similarly dressed, but without a hat. He is smoking a cigarette. On the wall behind him is a rack of cigarettes, a glass case filled with tapes and CDs, various pegs draped with fan belts, a shelf of sparkplugs, and a faded state highway map. Presently the boy at the Coke machine speaks without looking up.

MIKE
So whaddya think?

JEFF
(Looking up, confused.)
'Bout what?

MIKE
'Bout what we was talkin' about last night.

JEFF
Oh. Yeah.
(Pause.)
Be all right, if it worked.
(He looks down.)

                    MIKE
          (With conviction.)
It'll work.
          (Pause.)
Hey, Jeff.
          (The boy behind the counter looks up.)
Gimme a pack of cigarettes.

                    JEFF
          (He turns and surveys the rack.)
Whaddya want?

                    MIKE
I don't care.
          (Watches as JEFF pulls out a pack.)
Naw, not them.
          (JEFF's hand moves to another brand. He
          turns to get MIKE's approval)
Yeah, they'll do.
          (JEFF tosses him the pack.)
Say your boss never misses 'em?

                    JEFF
Naw. How could he? He'd never notice.

                    MIKE
No way he could.

                    JEFF
That's what I say.
          (They laugh. MIKE lights his cigarette,
          walks to the door and looks out.)

                    MIKE
Boss a pretty good fella?

                    JEFF
Yeah, he's pretty good. Easy to get along with.

                    MIKE
That's good.
          (He smokes, takes a swig of Coke, walks
          toward the counter, leans across it, and
          looks at JEFF.)
How much you reckon ya got in there?

                    JEFF
I dunno. Couple a hundred dollars. Probably not that
much. I dunno.

                              MIKE
                    (Lets out a low whistle.)
          What time does your boss come by to lock up?

                              JEFF
          'Bout twelve. Sometimes before.

                              MIKE
          What time is it now?

                              JEFF
                    (Looks at his watch.)
          Goin' on eleven.

                              MIKE
                    (After a long pause, looking at JEFF and
                    smiling.)
          Well, whaddya think?

                              JEFF
          I dunno, Mike. It'd be nice all right, but ya think
          it'd work?

                              MIKE
          Now's as good a time as any.

                              JEFF
          Yeah.

                              MIKE
          Might as well do it now, don't ya think? Wait any
          later and your boss might come by to close the place
          up.

                              JEFF
          Yeah, I reckon so.

                              MIKE
                    (Walks to the door and looks both ways at
                    the road.)
          Nobody else is gonna be comin' in now.

                              JEFF
          No.

                              MIKE
                    (Looking at Jeff.)
          Well, all right then.

                              JEFF
                    (Inhales deeply.)
          Okay. Who do ya wanna say done it?

                    MIKE
We won't say anybody. We'll just make up somebody

                    JEFF
Yeah, I know that, but what do they look like and all?

                    MIKE
I dunno. What sounds good?

                    JEFF
Wanna say some skinhead done it?

                    MIKE
Yeah. That's good. They'll believe that.

                    JEFF
Tall or short?
            (MIKE nods.)
Which one?

                    MIKE
Tall. But not too tall. Nothing unusual. Six foot, six
one. Something around there.

                    JEFF
How 'bout weight?

                    MIKE
Oh, about two hundred. Yeah, that's good. Sounds about
right.

                    JEFF
What should he be wearing?

                    MIKE
I dunno. Just regular clothes. Don't get into too much
detail, or it'll start soundin' funny. Tell 'em you
didn't get a good look at him, it happened so quick
and you was scared.

                    JEFF
Yeah. How about a mask?

                    MIKE
            (Looking out the door.)
All right.

                    JEFF
Anything else? How about a car?

                    MIKE
Naw. Then ya gotta start givin' models and colors and
all that crap. Just say he was on foot and leave it at
that.

                              JEFF
Okay. What should I do afterwards? Meet you someplace?

                              MIKE
Shoot no! It'll probably take ya awhile talkin' to 'em
anyway. I'll meet ya someplace tomorrow. I'll go over
to your house.

                              JEFF
Okay.

                              MIKE
               (Turns from door and walks to register.)
All right. Gimme a paper bag.
                    (JEFF pulls one from beneath the counter
                    and shakes it open.)
That looks good.

                              JEFF
Whaddya wanna bag for?

                              MIKE
Might wanna stash it someplace, just in case.

                              JEFF
Oh.

                              MIKE
               (Looks at JEFF.)
Go ahead.
                    (JEFF hesitates.)
I'll watch out.
                    (MIKE returns to the door and looks out.)
It's okay. Go ahead.
                    (JEFF takes money out of the register and
                    deposits it in the bag.)

                              JEFF
Coins, too?

                            · MIKE
Yeah, why not.

                              JEFF
               (Finishing.)
Okay, I got it.

                              MIKE
               (Takes bag and stuffs it under his coat.)
Okay. Call 'em. I'll leave soon as you call.

                              JEFF
               (Looking through phone book.)
When you gonna come over tomorrow?

                              MIKE
I dunno. Afternoon probably.
                    (JEFF starts to dial.)
Sound kinda scared when you talk to 'em.

                              JEFF
How scared?
                    (Voice shaking.)
Like this?

                              MIKE
No, not that much. Just choke now and then. It'll sound
all right.

                              JEFF
                    (Into phone.)
Ah, yeah. I'd like to report a robbery.

                    (MIKE smiles, nods at JEFF, waves, and
                    quickly exits.)

                         THE END

EVALUATION

In this subtle battle, Mike wants to convince Jeff to rob the service station, and eventually he succeeds. The objectives are clear and in opposition. In one sense the action here is psychological and internal—Jeff's decision to go along with the robbery. But the author also makes the action physical in that Mike both intimidates and encourages Jeff into the decision, which is the crucial moment of the piece. There follows, of course, the outcome, the physical action of getting the money, but that is simply finishing the scene. The conflict has been decided before that even starts.

Virtually all the dialogue arises from and is focused on the action of the scene. The characters speak because they want things or they have to respond. And their distinct personalities emerge through their interaction. The author of this scene has listened and observed carefully. The dialogue is realistic yet distinctive. The phrase that sparked the title—"Be all right, if it worked"—includes both a statement and a qualifier that makes a kind of crazy sense. The repetition in "'Bout what we was talkin' about last night" is similarly expressive. The author is also aware that when Mike says "I don't care" which cigarettes he gets, he actually does. He has also incorporated nonverbal cues. For example, Jeff turns to get Mike's approval for the cigarettes without actually saying anything.

The characters share a vocabulary, but they are clearly distinguished. Mike is the dominant force. He initiates and directs the conversation about the robbery, the cigarettes, the boss, and the money. Jeff responds only in bits and pieces. Jeff, however, is much more concerned than Mike with the details of the crime: Who should he say did it? What does he look like? What kind of car does he drive? When and where will they meet? Still, even in those concerns Jeff looks to Mike for guidance.

In this scene the author has also noticed important aspects of the setting. This is not a modern, high-tech convenience store. The dim inside light and the light from the outside sign suggest a shadowy mood.

Do we care about the characters? Insofar as we have all been subject to petty temptations of greed, probably so. Certainly the author succeeds in getting us interested in whether they will take the money and whether they will get away with it if they do.

The author is also successful in foreshadowing a future reversal in which the inexperienced Jeff will not see the money again. After all, if Mike fails to come by Jeff's house the next afternoon, Jeff can't really cry "foul" very loudly. And if Jeff should weaken and confess, Mike, with the money carefully hidden away, can simply deny any involvement whatsoever.

We have seen how any two characters can fall into direct conflict. We have examined how characters emerge through action and conflict and how characters express themselves in dialogue that springs from their objectives and the way they go after them. Now let's look at the wild complications that ensue when you add an extra ingredient to the mix—a third character.

5

· · · · · · · ·

Three-Character
Conflict

EXERCISE 5 • THREE-CHARACTER CONFLICT
Write a scene that places three characters in a situation of
modulated conflict. Write the scene using correct stage
terminology, with dialogue as needed, and bring the scene to a
resolution. Minimum length: 5–6 pages.

Adding a third character to conflict is like adding a joker to a card
game. The possible outcomes multiply dramatically. Mathematics plays a
part. Let's say you have two characters (for our purposes, James and Trish),
only four basic configurations are possible to express their feelings toward
each other.

1. They like each other.
2. They dislike each other.
3. Trish likes James, but James dislikes Trish.
4. James likes Trish, but Trish dislikes James.

James and Trish, of course, can be unique personalities and their
relationship subject to infinite degrees of passion, but the relationship must
eventually resolve itself into one of those four patterns.

When a third character is added (let's say, Hank), the number of
possible dramatic structures skyrockets. The four relationships listed above
are possible between Hank and James, and four more are possible between
Hank and Trish, in addition to the original four between James and Trish.
Instead of four possible relationships, you now have four times four times
four possible relationships. Sixty-four combinations instead of just four with
the simple introduction of a single character!

This point is not intended to make writing a play into an exercise in
mathematics. The numbers merely suggest the incredible complications that
can ensue when a playwright moves from two characters to three.

We read in theater histories that when Aeschylus wrote the first
Greek tragedies some 2500 years ago, he used only scenes of direct confrontation between two characters. Sophocles, whose career overlapped

Aeschylus, is credited with the introduction of a third character into scenes. Perhaps only a playwright can fully appreciate that contribution, for, as we will see, the third character opens a mine of possibilities.

The three-person conflict exercise in this chapter is an extension, but an important extension, of the second exercise. If conflict is basic to drama, then playwrights had better learn how to handle conflict. When two people are involved, the conflict is generally simple and direct: A is against B and B is against A.

As soon as a third party is introduced, however, the situation becomes more complex. Think, for instance, of two teen-aged brothers arguing about driving the family car. A direct conflict. Then the father walks in. Consider the possibilities. Each boy pleads his case, and then the father decides. Perhaps the father refuses the car to both. Perhaps the boys attempt to cover up their dispute in their father's presence in order to present a united front. Perhaps the father has a special fondness or a particular dislike for one son or the other. A good playwright will be able to use the various possible configurations and shift effectively from one to another.

As the conflict unfolds, the characters naturally reveal themselves through their actions, their words, their tactics, and their arguments. It is precisely that situation of a father and two sons that Arthur Miller employs so expertly in *Death of a Salesman*. Although his characters aren't just arguing about a car, Miller uses all the strategies mentioned above plus many more in the course of his drama.

As another possible three-character situation, imagine a young couple having a lovers' spat. Then the young man's mother enters. The choices for modulating the conflict are extensive. We could envision the mother supporting the woman who is trying to straighten out her son or supporting her son because he's her child. The couple might unite in telling the meddling mother the fight is none of her business.

As you can see from those hypothetical brothers and lovers, the three-person situation allows the playwright opportunities that do not exist in a two-person scene. A two-person conflict relies on a direct transaction. That is its great strength. The scene may contain a panoply of tactics and psychological ploys, but the action always has an essential one-on-one component.

THREE IN A ROOM

When a third person appears, an important indirect element arises. This example illustrates what I mean. A diamond is on a table, and two people are standing in the room with it. Suddenly the lights go out, and when they come on again, the diamond is gone. Because of the one-to-one relationship, both parties in the room must know who took the diamond. The one who took it certainly knows, and the one who didn't take it must know the other person did.

But, you suggest, couldn't someone else have come in when the lights were out and taken the diamond? Aha! You've just discovered the importance of the third character! With three characters in the room, the majority of the characters cannot know with certainty who has the diamond. Only the thief knows for sure. Now suspicion is possible; alliances are possible; deception and pretense are possible; majorities and minorities are possible. In short, conflict that is indirect and variable—or what I call *modulated*—replaces direct conflict as the primary form of interaction.

So important is the three-character situation that certain three-party scenes have become standard dramatic fare. With infinite variations they are used over and over again.

A trial is a typical three-party modulated conflict. Two parties modulate their direct conflict by attempting to convince a powerful third party to decide in their favor. That is the essential configuration in every courtroom drama ever written. The complainant and the defendant are in opposition, but instead of fighting with each other directly, both are trying to make their case to the third party, represented by the judge or jury. Although more than three characters are usually involved, the conflict is essentially triangular.

The same "trial" structure, which is basically an appeal to authority to make a decision, forms the basis of many other conflicts. Two young men contend for the love of a woman, and they appeal to her to decide between them. Siblings vie for the attention of a parent, each pleading, in effect, "Say you love me more."

A slight variation of the trial format is the peacekeeper. In the example of the trial, two contending parties appeal to the third party to make a decision. The peacekeeper steps between contending parties in an attempt to separate them. The peacekeeper may or may not mediate the conflict, but the immediate goal is to prevent direct conflict. Such is the position of a police officer who breaks up a fight between two thugs.

The next logical step in the three-sided situation has two of the parties joining forces against the third. The judge hands down a verdict. The girl chooses her beau. The point at which such a decision is made is crucial for a playwright, for once two of the parties (A and B) join forces against the third (C), the situation is for all practical purposes changed to a direct conflict (A/B vs. C). Hence good writers try to use the pairing of forces in original ways. Perhaps the audience will expect B to join A, but B joins with C instead. Perhaps A and B oppose C on one issue in the play, but B and C oppose A on another issue, and A and C oppose B on yet a third issue. Through creative pairing of forces, a playwright can produce a rich interplay of shifting alliances.

Another common three-sided situation is a circular one. A wants to gain B's support in opposing C, while B wants to gain C's support in opposing A, and C wants to gain A's support in opposing B. Or the same circle can be created in terms of desire. A wants B, who wants C, who wants A. That circle of desire joined with an equal circle of hate comprises the classic "hellish" situation created by Jean Paul Sartre in *No Exit*.

Another frequently used three-way situation develops when at least one character is not aware of the presence of a third party. Such is the case when one character is hidden or is listening in on a conversation. That arrangement has produced some of the most famous scenes in all of drama, including Hamlet's emotional scenes with Ophelia and Gertrude, when Polonius is an unseen listener.

A scene closely allied to the hidden character is that of disguise, in which at least one of the characters does not know the true identity of another character. Even if a scene of mistaken identity contains only two characters, it assumes a third character. If I think you are Elvis Presley returned to life, then our interchange contains you, me, and the assumption of Elvis. Such mistaken identity scenes have been a cornerstone of comedy at least since Plautus wrote *The Twin Menaechmi*, in which identical twins are regularly confused one for the other. Shakespeare adopted the idea for *Twelfth Night* and *The Comedy of Errors*. More recently, Ken Ludwig used similar tactics in his comic hit, *Lend Me a Tenor*.

DRAMATIC IRONY AND LOVE TRIANGLES

Scenes of hidden and disguised characters both exemplify what is called *dramatic irony*. Dramatic irony occurs when the audience knows something that is unknown to a character. The audience, for example, knows that Viola has disguised herself as a male in *Twelfth Night*, but the characters who fall in love with "him" do not. The audience knows there is a set of twins in the Plautus play and two sets in Shakespeare's adaptation, although none of the characters realize it.

Finally there are the *love triangle* situations. A love triangle does not refer to one distinct relationship among three people. Rather, there are numerous variations, all of which qualify under the rubric "love triangle." I have already mentioned one love triangle in discussing the "trial" format—two men both love the same woman, and they appeal to her to make a decision—and another in the circular format where A loves B who loves C who loves A.

Another love triangle might involve a husband or wife who loves his or her mate, but the mate loves someone else. You could have someone who is loved by two individuals who may or may not know each other. The possibilities are numerous enough to have allowed inventive playwrights over several centuries some measure of fame and fortune. Many modern playwrights have given new life to the standard triangles by altering the usual gender of the participants, as Terrence McNally did in *Love! Valor! Compassion!* and Tony Kushner did in *Angels in America*. The essential triangular love relationships, however, remain the same.

As you undertake the three-person exercise described in this chapter, remember that you can learn not only from what *you* write, but from what *others* write, as well. If you are in a classroom situation, pay careful attention to the differences in the three-person scenes. Consider how the relationships between the characters could be altered or rearranged. If you are not in a class, give particular consideration to three-person scenes in the plays, movies, and television shows you see. The nurse, the patient, and the visitor in *The English Patient*, for instance, lead to a complex set of triangles.

Once a writer has learned to handle two characters in a direct conflict and three characters in a modulated conflict, he or she can confidently tackle almost anything. Additional characters or new alliances merely present variations based on direct or modulated conflict. As an example of what I mean, recall the lovers who were arguing. Assume they're fighting because she wants to move in with him, but he doesn't want her to. Assume the young girl's parents enter, and the boy and girl tell them what's wrong. The father angrily rejects the idea that his daughter should live with someone before marriage. The mother tries to soothe her husband's anger toward their daughter, but when he stands firm, the mother takes her daughter's side and announces that she wishes she had lived with her husband for awhile before she decided to marry him. Let's examine how such a scene might be structured.

Scene action	*Scene structure*
1. The young man (A) and young woman (B) argue	1. Two-party direct conflict: A vs. B
2. The woman's parents (C and D) enter and the daughter and boyfriend appeal to them for support	2. Three-party modulated conflict ("trial" format): A vs. B, both appealing to C/D
3. The father sides with the boyfriend	3. Two-party direct conflict: A/C vs. B
4. The mother tries to mediate between father and daughter	4. Three-party modulated conflict (mother as "peacekeeper"): D mediates between A/C vs. B
5. Father rejects mother's mediation attempts	5. Two-party direct conflict: D vs. C
6. Mother supports daughter and announces that she wishes she'd done it	6. Two-party direct conflict: B/D vs. A/C
7. Father, stung, accuses wife of not loving him	7. Two-party direct conflict: C vs. D
8. Young couple join in attempt to end strife between mother and father	8. Three-party modulated conflict (young couple as "peacekeeper"): A/B between C vs. D
9. Mother and father storm out; young couple resolve to set the disagreement aside and discuss it later	9. The scene, even though problems remain, is resolved

The two-party conflict that begins the scene ends as the young couple join forces to stop the fight between the woman's parents and then decide to temporarily set aside their differences. Notice that at no time is there actually a four-way conflict. I would not go so far as to say that four-way conflict is impossible, but I would assert that, as parties proliferate, they usually team up to create combinations of direct two-party or modulated three-party conflicts.

• • • EXERCISE 5.1

Write a scene that places three characters in a situation of modulated conflict. Write the scene using correct stage terminology, with dialogue as needed, and bring the scene to a resolution. Minimum length: 5–6 pages.

The factor that distinguishes this exercise from the conflict exercise in Chapter 2 is, naturally, the presence of a third character. In evaluating the scenes you write, ask yourself some crucial questions: How does the presence of three people affect the conflict? Can you identify what each character wants and how he or she goes about getting it?

Keep in mind also the fundamentals of the previous exercises. What action does the audience see taking place? Does the dialogue flow naturally from one character to another? What kind of language does each character use?

You should write a completely new scene with three characters, but after you have done that you may wish to add a third character to one of your previous two-person scenes to see how that changes the interaction.

To understand how the addition of a third character affects a scene, let's examine a student-written piece that begins with two characters and then adds a third.

EXAMPLE

BREAK
By Debbie Laumand

>   THE SETTING is an ordinary living room. Mom
>   is seated in a chair, reading. BETH, age
>   21, ENTERS. She is carrying a book.

                         BETH
I thought I'd find you in here reading.

                         MOM
            (Not paying attention; keeps reading.)
Yes, dear.

                         BETH
I like to read in here. It's nice and quiet.

                         MOM
            (Nods head vaguely.)
Um-hum.

                         BETH
Nobody to bother you. You know what I mean?

                    (MOM nods again. BETH makes a horrid face
                    at her mother to see if she's paying atten-
                    tion. She isn't.)
I'm gonna read some Shakespeare, Mom.

                              MOM
                    (Nodding.)
Um-hum.

                              BETH
You know what my English professor says about
Shakespeare?

                              MOM
Um-hum.

                              BETH
He says if I don't read Shakespeare, I'll become frigid
before I'm twenty-five.

                              MOM
That's nice, dear.

                              BETH
Yep, Mom. That's right. The ol' "spread the legs and a
little light turns on" routine. No Will, no will. You
know what I mean? Ice cubes, Mom. I'm talking crushed
ice.

                              MOM
That's nice, dear.

                              BETH
                    (Sighs.)
I guess I'll read now.
                    (BETH sits and begins to read. A door open-
                    ing and closing is heard offstage and a
                    light infiltrates, as if someone has turned
                    on a light in another room.)
I guess Ben's home.

                              MOM
Um-hum.
                    (A TV set is heard from the next room.
                    Loud. A football game is on.)

                              BETH
God! There go the holiday football games. Does he turn
that dumb thing on the second he walks in the door?
                    (No reaction from MOM. BETH gets up and
                    goes to the entranceway. She yells at BEN.)
Must you subject Mother and myself to that mindless
drone at decibel rates far beyond the endurance of the

human ear? Is it absolutely necessary to have the sound
so loud that one's hair is parted by the sheer force
of it? You may not understand this, Ben, but mother and
I are practicing an ancient art called "reading." We
are expanding our minds with the written word as opposed to shrinking it with an array of unconnected
molecules, more commonly known as the "television image." And we don't appreciate your blatant and rude
behavior in turning . . .

                    BEN (Off)
Aw, shut up, Beth. You're fulla shit.

                    BETH
You watch your language. You may be fifteen now, but
that doesn't mean you can speak to me that way.

                    BEN (Off)
Stuff it!

                    BETH
I am merely trying to express my thoughts in the most
civilized manner possible. I am trying to reach you
through the spoken language. I am asking you nicely to
turn the volume down.

                    BEN (Off)
Go to hell!

                    BETH
You turn that volume down or I'm gonna come in there
and beat the crap out of you, and I'm still big enough
to do it, bucko!
            (The volume is turned down. MOM looks up at
            BETH. BETH catches the look. MOM looks down
            and resumes reading. BETH crosses to her
            chair and picks up her book.)
The only language these heathens understand is violence. I don't believe him. He didn't act like that
when I lived here, did he? I mean, what has happened
to this family?
            (No response. BETH begins to read again.
            Suddenly BEN is heard screaming rapidly
            from the other room.)

                    BEN (Off)
Alright. Alright! He's got the ball! Run, you sucker
run! Do it! Do it, man! Do it! Do it! Do it! Yeah!
Yeah! Yeah! Whoa!

                    BETH
            (Starting, eyes wide.)
Good God! What is he screaming about? Does he always do
that?

                    MOM
            (Not looking up.)
Yes, dear.

                    BETH
Jeez, my heart's pounding like crazy.
            (No response from MOM. BETH returns to her
            book.)

                 BEN (Off)
What the...! What kinda ref are you? Are you nuts?
Aw, what a bummer!

                    BETH
            (Starts, looks up, and yells.)
Knock it off, Ben!

                 BEN (Off)
            (Loud.)
Oh, man! What a catch! Go with it, man. Alright! Go!
Go! Do it! Do it! Do it! Alright! Wow!

                    BETH
            (Looks down at book and overlaps LOUD as
            BEN continues to yell.)
Oh, man! What a bummer! The poison won't kill Juliet.
But wait. She's going for the knife! Alright! She's
gonna stab herself. Alright! Do it! Do it do it do it
do it! Yeah! Alright! Way to go, Juliet!

                    BEN
            (At entranceway.)
Beth, you're a real hag, you know that?

                    BETH
            (Goes to him.)
Listen, twerp, I've had just about enough of you!

                    BEN
Yeah? Well why don't you do something about it?

                    BETH
Alright jerko...you better believe I'm...

                    MOM
            (A forceful statement, but not a yell.)
That will be quite enough, you two.

                              BETH
But he—

                              MOM
Beth, quiet. Ben, you watch your language and stop
provoking your sister.

                              BEN
Yes, ma'am.
                    (BEN returns to the other room.)

                              BETH
Good. You really have to keep a strong control over
these kids, or they just—

                              MOM
And you, young lady . . .

                              BETH
Ma'am?

                              MOM
You stop provoking your brother.

                              BETH
He was the one with the screaming problem.

                              MOM
Beth, he lives in this house just like you do, and he
has a right to be here, just like you do.

                              BETH
But he was disturbing my reading. How could I—

                              MOM
My! My! My! Me! Me! Me! You've been a tyrant since day
one of your vacation.

                              BETH
I have not. . . .

                              MOM
Yes you have. I don't know what you're going through
right now, Beth, but it has got to stop. You were
treating your brother as if he didn't belong here.

                              BETH
Everybody's changed. I don't feel like this is my house
anymore. I was so looking forward to the Christmas
break, Mom. I just wanted it all to be like . . . well,
like a Hallmark card.

                              MOM
Beth, this is your home. It's the same home it's always

been, and we love you. And if you don't straighten up, "I'm gonna beat the crap out of you, and I'm still big enough to do it, bucko!"

>   BETH
>
>   (Laughs.)
> I'm sorry, Mom.

>   MOM
> That's okay. I'm going into the kitchen to get myself a glass of tea. Do you want any?

>   BETH
> Yeah. I'll go with you.

>   MOM
> Good. But leave your Shakespeare book here, darling. The freezer's broken, and I want to see you make some crushed ice.

>   BETH
> Mom!
>   (THEY EXIT.)

>   THE END

EVALUATION

The basic conflict of this scene is between a young woman who wants attention and her mother, who thinks her daughter is being too self-centered. The playwright has an inherent problem to solve. Since the mother is ignoring the daughter, the scene seems to have nowhere to go. Until the introduction of the third character. The arrival of Ben in the next room provides Beth with an opening, an excuse to create a disturbance.

At its core the scene has the structure of a trial situation. Two siblings are contending for the approval of a parent. What makes the scene unusual is that the conflict between the siblings is deemphasized in order to increase the attention on the relationship between the girl and her mother.

The writer makes an intriguing choice in keeping the third character almost entirely offstage. Normally that would not be a wise choice, and the image of the girl standing at the doorway talking to an offstage character might seem contrived. How would it have affected the scene if Ben were on stage? Naturally Ben would have become a much more prominent figure. And the direct conflict between Beth and Ben would become the central element of the scene. By keeping Ben offstage, the focus remains on Beth and her mother.

Are the characters differentiated in the way they speak? Beth conveys distinctive qualities through her language. Her speech has a broken, jagged quality that suggests youth and quickness: "No Will, no will. You know what I mean? Ice cubes, Mom. I'm talking crushed ice." Her lines reveal not only that she is educated, but that she is bright and witty. That information is not conveyed through direct means. No one, thank heavens, says, "My what a bright and witty girl you are." We don't conclude that she's intelligent because she's in college or because she's reading Shakespeare. Lots of not particularly bright individuals go to college, and some even read Shakespeare. What convinces us that Beth has a good mind is in part her ability to handle language: "Must you subject Mother and myself to that mindless drone at decibel rates far beyond the endurance of the human ear?" But even more, we are convinced of her intelligence by her sense of style, by her ability to shift from overblown verbiage to common language in a trice, as when she moves from "I am merely trying to express my thoughts in the most civilized manner possible . . ." to "You turn that volume down or I'm gonna come in there and beat the crap out of you . . ." in the space of a single line.

Beth's consciously elaborate language and her mocking imitation of her brother's speech reveals that she is self-dramatizing, and in the drama that she creates she is also quite entertaining. That, after all, is a must for plays.

The audience finds out much less about the other two characters through the language they use. Ben talks in slang at virtually every line: "What a bummer" and "You're a real hag." From that we get a sense of what Beth is attempting to place herself above. Still, through her calculated use of similar slang we understand that it remains a basic part of her.

The mother has few lines, yet they disclose her character fully. Her repetition of "Um-hum" and "That's nice, dear" indicates someone who is not paying attention and suggests that Mom is not particularly interested in her daughter. We soon find, however, that Mom is in charge in her home. She effectively stifles the spat between Beth and Ben, and her direct, plain talk to Beth reveals her as both forthright and caring. Finally, her last lines demonstrate that she heard and cared about everything her daughter said.

Now let's look at a tight, complex scene that uses three characters throughout.

EXAMPLE

NIBBLES
By Mary Kerr

THE SETTING is a living room. There is a couch center, facing the audience. A small portable TV sits on a low table down center in front of the couch. MIKE and DREW sit on opposite ends of the couch watching TV. Both men are about 20. STEPHANIE, about the same age, ENTERS carrying a bowl of chocolate chip cookies. She sits between the two men.

                    STEPHANIE
I made some cookies to eat while we watch the game.
They're still hot.

                    MIKE
Alright!
            (He grabs a handful of cookies out of the
            bowl and begins eating.)

                    STEPHANIE
            (Offering bowl.)
Drew?

                    DREW
Oh, no thanks.

                    STEPHANIE
You're sure?

                    DREW
Yeah. Thanks anyway.

                    STEPHANIE
Come on. They're real good.

                    DREW
I don't want any, Steph, really.

                    STEPHANIE
Here I go to all this trouble, and you're not going to
eat any?

                    MIKE
Drew's on a diet.

                    STEPHANIE
What? Drew, is that true?

                    DREW
Yeah.

                    STEPHANIE
But that's silly. You're not fat.

                    DREW
I'd like to drop about five pounds.

                    STEPHANIE
That's crazy. You look fine. Come on, have a cookie.

                    MIKE
He doesn't want one.

                    STEPHANIE
One cookie won't hurt him, Michael.

                    MIKE
Look, Steph, just leave him alone, okay?
          (MIKE takes four more cookies from the
          bowl.)

                  STEPHANIE
Don't eat 'em all. Save some for Drew.

                    DREW
I really don't want any.

                  STEPHANIE
But you aren't fat!

                    MIKE
Jeez, the guy is a little overweight and he's trying
to do something about it, and you won't leave him
alone.

                    DREW
Hey, yesterday you said you didn't think I needed to
lose any weight.

                    MIKE
Well...ah...I just meant that you thought you were
overweight. I don't think you're overweight.

                  STEPHANIE
Neither do I, Drew. So have a cookie.

                    DREW
Don't tempt me, Steph. I really don't want one.

                    MIKE
Good! That's more for me.
          (He reaches into the bowl for more
          cookies.)

                  STEPHANIE
Mike, you could afford to forego a few cookies
yourself.

                    MIKE
Oh, really?

                  STEPHANIE
Yes, really. You're sprouting some pretty hefty love
handles.

                    MIKE
Yeah? What about you? Those jeans weren't that tight
when you bought 'em.

                              STEPHANIE
Am I getting fat?

                              MIKE
I didn't say fat. Your butt's just gettin' a little
wider is all.

                              DREW
Don't listen to him, Steph. Your butt looks just fine to
me.

                              MIKE
When were you lookin' at her butt?

                              DREW
Jeez, Mike, she lives here. I see her butt every day.
It's not like I've been staring at it.

                              STEPHANIE
            (Standing up, looking down at her rear.)
Does it really look alright?

                              DREW
It looks fine. Your whole body looks fine.

                              STEPHANIE
So does yours.

                              DREW
Thanks.
            (STEPHANIE and DREW smile at each other as
            she sits back down and they resume watching
            TV.)

                              MIKE
There's three cookies left. Drew?

                              DREW
No, thanks.

                              MIKE
Steph?

                              STEPHANIE
No thanks. I think I've had enough. Go ahead, Mike, you
finish them off.

                              MIKE
No, uh, I'm not hungry anymore. We can save 'em for
Kurt for when he gets off.

                              STEPHANIE
That's a good idea. Kurt can eat them.

(They sit looking at the TV in silence.
STEPHANIE lights a cigarette. MIKE gets out
a piece of sugarless gum. DREW begins to
chew his fingernails.)

THE END

EVALUATION

On purely technical grounds this scene is an excellent example of what can be accomplished with a three-person scene built on shifting alliances. It begins with Stephanie (A) thrusting cookies upon Drew (B), who refuses (A vs. B). Mike (C) supports Drew's refusal (A vs. B/C). The conversation moves to the topic of weight. Drew maintains he's overweight, and Stephanie disagrees. At first Mike seems to agree with Drew, but in one of the nice ironies of the scene Mike is forced by the person whose side he is on to switch sides ("I just meant that you thought you were overweight. I don't think you're overweight"), thereby generating A/C vs. B. Stephanie immediately upsets that arrangement by insisting that Mike is overweight (A vs. C), and he retaliates by accusing Stephanie of heftiness. Drew, however, supports Stephanie, creating an A/B vs. C situation and bringing together the two people who were at odds at the beginning of the scene.

The scene is notable in a number of other ways. First, despite its limited physical action, it contains large psychological movements. Second, it produces those movements in a spare, concise manner. Third, it conveys a strong sense of sexual tension without any direct reference or overt action.

The physical activity of the scene is limited to three young people watching television and eating cookies. The conflict at the surface of the scene is Stephanie's desire for Drew to eat the cookies she's just baked and his desire, supported by Mike, not to eat any cookies.

That leads to a second level of conflict. Is Drew fat? Drew thinks so, Stephanie doesn't, and Mike's position shifts. Is Stephanie fat? Mike says so, but Stephanie and Drew say otherwise. Is Mike fat? Stephanie suggests he is, but Mike denies it.

Physically the scene is nearly static, with only the passing of the bowl and the munching of cookies. But psychologically the ground covered is vast. Mike goes from ease and confidence to insecurity and virtual denial of appetite. He begins as a friend of both Stephanie and Drew, but he winds up strangely alone. Drew shifts from alliance with Mike over his own need to lose weight to alliance with Stephanie. Stephanie, in effect, moves from Mike to Drew, and we see in small details the potential end of one romantic relationship and the beginning of another.

Psychologically, then, this scene is about sexual relationships, and it is to the author's credit that she has written a highly charged, erotic scene with hardly a word of sexually explicit language. Yet the sexual overtones are clearly there, in the focus on Drew's body, in the emphasis on Stephanie's "butt," in the reference to Mike's "love handles," and in every gluttonous mention of the chocolate chip cookies.

Although there is much in this scene to admire, there are also ways in which it might be improved. For one, we completely lose sight of the fact that they are watching television. Since that is, on the surface, a primary action, it ought to used. The scene would benefit from references to appropriate images on the screen—perhaps weight lifters, *Baywatch* hunks, or three-hundred-pound tackles.

A second primary action of the scene is eating the cookies, and there, too, more could be done with the action. Attention might productively be drawn to the ways that Stephanie and Mike eat their cookies. Perhaps Mike makes a mess, dropping crumbs on his shirt or crumbling pieces of cookie onto the sofa. Perhaps Stephanie eats her cookies slowly, nibbling around the chocolate chips, which become sticky on her fingers so that she licks off the wet chocolate. Precisely how to incorporate watching the television and eating the cookies more noticeably into the action is for the playwright to determine, but attention to those primary actions could add scope to the scene.

A second area in which the scene could be improved involves the use of language. The dialogue of the scene flows easily from one character to another, and it sounds natural. We can believe that people talk like that, and that is a solid virtue. What is lacking is clearer differentiation in language patterns between Mike, Drew, and Stephanie. In terms of the words used, and the way the words are put together, any of the lines could be spoken by any of the three characters.

Nibbles depends on there being three characters. Without Drew in the scene, Stephanie and Mike would simply have eaten the cookies without a word about weight and without conflict. Without Mike, Drew would simply have refused the cookies and would never have declared how appealing he thinks Stephanie looks. And, of course, without Stephanie and the cookies there is no scene. Such are the complications of the three-party scene.

Have you developed a sense for creating action? For showing your audience characters doing something? Can you put two characters into a direct conflict and manage the scene? Have you gained a sense of how people express themselves in language? Have you come to terms with the variety of ways three parties to a conflict can interact? If so, then you have the basic tools needed to prepare any scene in drama. In the following chapters you'll be able to use those tools to develop additional scenes that will bring your characters and your ideas to life.

6

· · · · · · · ·

Writing from Life

EXERCISE 6.1 • WRITING FROM LIFE
Write a scene based on something you know well from actual
life. Use stage terminology. Use as much dialogue and as many
characters as you need. Resolve the scene. Minimum length: 6–7
pages.

In many writing classes—not only playwriting but other forms of
creative writing as well—students are advised to write from their personal
experiences. That is an excellent place to start, and in this chapter you will
be asked to write a scene about something you know extremely well, a
scene that uses as its seed something taken from your own experience.

Every writer writes from his or her own experiences. In some cases
the dramatic creations are quite close to real life events in the writer's life.
Tennessee Williams used remembrances of his mother and invalid sister
to help him shape Amanda Wingfield and Laura in *The Glass Menagerie*.
Edward Albee's relationship with his mother provided the basis for *Three
Tall Women*. Ntozake Shange, who wrote *For Colored Girls Who Have Consid-
ered Suicide/When the Rainbow is Enuf*, has stated very candidly that everything she writes is founded in her race and her gender. Alfred Uhry used
memories of his own relatives in writing the delicate prize-winning plays
Driving Miss Daisy and *The Last Night of Ballyhoo*.

Eugene O'Neill's *A Long Day's Journey Into Night* examines family
relationships through characters that to some degree mirror O'Neill's own
family: a father who is a famous actor, a mother living in a drug-hazy past,
a wastrel brother, and the playwright's own unhealthy self. Even in his
earlier plays O'Neill's realistic seamen were based on men he had known
while laboring on sea-going cargo vessels.

Tina Howe's early experiences at the Metropolitan Museum of Art
influenced her writing of *Museum* and *Painting Churches*. Her own parents
bear a strong resemblance to the parents in the latter play. Unfulfilled plans
for a European excursion with her HIV positive brother inspired Paula
Vogel's *The Baltimore Waltz*.

Plays about political subjects as well as familial ones can materialize
from personal experience. Before Vaclav Havel became the president of the
Czech Republic, he endured years in communist Czechoslovakian prisons

as a dissident writer, and those bitter ordeals inform *Largo Desolato* and all his other plays.

Even when the subject appears to be distant from an author's actual life, the author is still using his personal experience. Famous men and women are frequent subjects of plays. Peter Shaffer's *Amadeus*, Stephen Sondheim's *Assassins*, Caryl Churchill's *Top Girls,* and August Wilson's *Ma Rainey's Black Bottom* are just four examples. But the playwright must always filter factual materials through the eyes of his or her own experience. Consequently, Shaffer shows us personal and artistic jealousies erupting between classical composers Salieri and Mozart through the glass of his observed artistic vanities. Churchill's historical characters reflect on women's issues with a twentieth-century consciousness, just as Sondheim's assassins call our attention to modern issues of politics, fame, and violence. Wilson's *Ma Rainey* is set in the 1920s, but it encompasses Wilson's own end-of-the-twentieth-century views on race relations.

In short, writers write from their own experience because they can't *not* write from their own experience. Who else's experience can they use? Who else's can *you* use?

Sometimes, particularly in dealing with historical material, a writer conscientiously attempts to use only documented facts. That may lead to good history, but it usually leads to bad plays. I don't mean that facts must be changed or distorted—although Shakespeare and Marlowe certainly had no hesitation about altering history to create better drama in their day, and neither does film director Oliver Stone in ours. But when you write a dramatic character, that character must come to life on stage. The character must act, talk, move, think. The dramatist cannot possibly know all the minutia about even a well-known historical figure, and so, as a result, the writer must fill in the details in order to bring that person to life. And those details must come from the writer's own background and be filtered through the writer's mind and feelings.

I am not suggesting that historical research plays no part in the creation of such plays. I'm sure that Shaffer and Churchill and Wilson and Sondheim studied their subjects carefully. What I am saying is that the research itself becomes part of the writer's experience—an ancient account, if you will, sifted through a contemporary consciousness—and that everything with which a person comes in contact, from personal observations to book learning, becomes a part of that individual's experience.

Writing from your own experience has several advantages. Because you are writing about places, people, and experiences that you know well, the writing frequently contains particulars that imbue the scene with depth and richness and give to the characters a set of very human idiosyncrasies.

EMOTIONAL CONNECTIONS AND HAZARDS

Another advantage in writing about something you know well is that you will usually have some emotional connection with the material. You may even care deeply about a person or a situation. If you can find a dramatic way to make the audience care deeply as well, you will have accomplished wonders.

This point of caring about your material can hardly be overstressed. In fact, many writers view it as a crucial starting point. Wendy Wasserstein once said that what drives her to write comes from experiences that she's actually had as a woman. You must care, and care deeply, about your material and what you want to show us if you hope to make anyone else care about it. After all, if you don't care about what you're writing, why should anyone else?

That's why a personal experience often provides such a good springboard to writing. Whether you're writing about a situation you've known or a person who has crossed your life, the chances are you will have a personal and emotional involvement. If the situation involves a family member, you will probably have strong feelings one way or another, just as O'Neill and Albee did. Possibly there will even be conflicting feelings of love and hate. All of that creates fertile ground for dramatic possibilities.

Just as there are advantages to writing from your own experience, there are also potential hazards. One danger lies in an overreliance on an actual situation. I recall an instance when a young woman wrote a very good play about a young woman breaking away from her family and learning to stand on her own. There were several excellent scenes involving the family. They showed that the family members loved each other but unconsciously stifled each other as well.

A flashy boy, who was essentially a negative character, played an instrumental role in the young girl's break from her family and its traditions. Most of the audience who heard a reading of the play thought the flashy boy was depicted as too evil and conniving. It did not seem dramatically plausible that the reasonably intelligent girl would have been taken in by him. That created a problem, especially since the boy served the positive dramatic function of helping the girl achieve her independence.

The author's response was: "That's how he was." And that's the danger. Something may happen in real life, a person may behave in a particular way, yet when those events are placed on a stage, they may not seem believable. It is inadequate for an author to say "That's how it really was" because, finally, the audience won't know or care how it really was. They will only know what they see on the stage, and that dramatic representation must seem plausible for them to accept the characters and their actions. It is not always possible to predict what an audience will find plausible and what it will reject. Generally, dramatic plausibility means that the audience can accept that the characters could feel the way they are shown to feel or behave the way they are shown to behave.

In *Othello*, Shakespeare presents the audience with a strong, active, military man. He is a Moor in a white society. He is married to a white woman whose father opposed the marriage. It is plausible that, like many a newly married man, Othello may fear the loss of his beautiful wife's affections. When he begins to grow jealous, he tests her—a plausible action. When his reasons for suspicion and jealousy are apparently confirmed, it becomes plausible that this strong man could, out of a sense of betrayal, destroy the thing he loves most in the world.

Playwrights, then, must never forget their first responsibility is to the play. The playwright is not a journalist who must recount each detail correctly. Rather, the playwright is an explorer, imagining why people act as they do and creating fictional, dramatic figures that must seem right and plausible. As you write, you might keep in mind what Tina Howe likes to say about her plays: "All the events are true, but none of them ever happened."

A second hazard, somewhat related to the first, is that of becoming overindulgent with your own interests. When you know something intimately, it is easy to give too much detail about a room, a character, or an event. Only those elements that are dramatically necessary to the action or to the characters need to be included. It is also easy, because you know the situation so well, to provide details that, to you, have obvious connections but seem extraneous to an audience.

What happens if you dramatize something that you think is incredibly interesting, but that bores the audience? It's not the audience's fault. You, as the playwright, have failed in some way. Perhaps you have not made clear to the audience the connections you perceive. Perhaps you have simply failed to translate what you found interesting in those connections into something that is dramatically interesting in the world of the play.

Yet another danger of writing based on personal experiences is that of violating other people's privacy. Friends or family members who are involved in an incident that you have used as the basis for a dramatic scene may recognize themselves and not be entirely pleased. Worse yet, other people may recognize them. Bear in mind that when you use personal material as the basis for a scene, you are not attempting to recreate reality. You are attempting to construct an entertaining, moving, and perhaps meaningful and instructive dramatic work. To that end, it is advisable to provide some distance by changing names, dates, places, and so forth. More importantly, it might be dramatically necessary to alter events, characters, and motivations to satisfy the particular needs of your dramatic work.

The final danger of writing based on personal experience is a psychological one. Sometimes those things in our lives that are inherently the most dramatic are also the most threatening or the most disturbing. When Marsha Norman wrote her Pulitzer Prize–winning play '*night Mother* about a woman who tells her mother she is going to commit suicide, she called it a highly personal piece. All writers enter such terrain with caution. On the positive side, writing from personal experience frequently allows you to reflect on situations and to understand better both the situation and your own personality and behavior. It can, in other words, help you to "know yourself" and, in so doing, make you a more honest and sensitive writer and person.

Nevertheless, that very process of self-discovery can be difficult, even painful. If in examining material to write about you discover areas that are troubling, you have three choices. One is to write about something else. A second is to pursue the area, while remaining aware that it may cause anxiety, stress, or even more serious problems. I don't wish to sensationalize, but a young writer should be aware that such exploration has led both to excellent plays and to some tortured, damaged lives. A third choice is to seek professional counseling for the problem. That may or may not improve your playwriting, but it might help your adjustment to life, which could ultimately be more beneficial, anyway.

A final question arises regarding the use of this exercise with young writers: How can young people, who by and large have not had wide-ranging personal experiences, have adequate resources on which to draw? That isn't really so great a problem as it seems. True, the experiences of many young writers are limited to those of a student in a classroom and a child in a family. True, as well, that a group of students using this exercise will more than likely turn up a fair share of plays about school life. But that's all right. Material does not have to be original to be treated originally. In fact, the best material is usually something with which everyone is familiar. There are thousands of plays about family life, about young people coming of age, and about young love. It isn't important that a student write about something that has never been written about before. The student who tries that will probably give up writing in despair. What *is* important is that each student discovers his or her unique viewpoint, a special way of analyzing, understanding, and dramatizing a situation.

Furthermore, don't underestimate the breadth of experience of even a relatively young person. Family and school situations comprise two of the most obvious starting places, but there are numerous other sources: friends, travel, camps, sports, other extracurricular activities, unusual incidents, and jobs are only a few.

With those explanations and warnings, let's look at the exercise and two student-written examples.

• • • **EXERCISE 6.1**

Write a scene based on something you know well from actual life. Use stage terminology. Use as much dialogue and as many characters as you need. Resolve the scene. Minimum length: 6–7 pages.

You may find as you work within this format that you have a lot to write about. Because you know the situation well, you will have a great deal of material from which to draw. You could even find yourself writing what you think is too much. That's all right. Don't worry about it. Just get down on paper everything you want to write. You can be more selective, if that's needed, in revisions.

Be aware also of the elements you've worked on in previous exercises. What actions do we see occurring? What are the conflicts? Is it direct, between just two people, or is it modulated through the presence of one or more other characters?

The following scene used an incident from the author's experience as a springboard to a two-character scene with rather indirect conflict.

EXAMPLE

CLICHÉ

By Alex Domeyko

THE SETTING is a city sidewalk at 3 A.M. Bits and pieces of trash litter the sidewalk, indicating that this is not the best part of town. It's a cold night. CATCH, a street bum, is in his early 30s but looks and acts older. He is sitting on the sidewalk leaning against a building, drinking a cup of coffee. CHRIS, a young man in his late teens, ENTERS, walking slowly along the sidewalk. He notices CATCH, stops, pulls some change out of his pocket, and drops it into his cup of coffee.

CATCH

Hey, kid. I was drinking that.

CHRIS

Oh, jeez. Sorry. I didn't know there was something in it. I thought you wanted change.

CATCH

Uh-uh, kid. I'm off duty.

CHRIS

Well . . . uh . . . keep it anyway. Sorry.
      (Begins to walk away.)

CATCH

Hey kid. I said I'm off duty.

CHRIS

      (Stops, turns around and faces CATCH.)
Well . . . I just thought you might want it anyway. I mean, don't you need it?

CATCH

What?

CHRIS

I just thought. . . .

CATCH

Look, kid. I don't need some little diaper bag telling me what I do and do not need, okay? I run a business here, and right now I'm closed. Okay?

                         CHRIS
A business?

                         CATCH
That's right. A business.

                         CHRIS
What kinda business?

                         CATCH
Good will. You got a problem with that?

                         CHRIS
No, no. Not at all.
         (Pauses to see if there's more coming.)
Well . . . sorry to bother you.
         (Begins to walk away, then hesitates and
         turns around. He starts to say something,
         but then turns to walk away again.)

                         CATCH
Catch.

                         CHRIS
         (Stops and turns around again.)
Catch what?

                         CATCH
Catch. That's my name.

                         CHRIS
Oh. My name's Chris.
         (Searching for something to say.)
Why do they call you Catch?

                         CATCH
You see how my nose is all crooked?

                         CHRIS
         (Getting closer to look.)
Yeah.

                         CATCH
When I was a kid, my brother threw a baseball at me.
Yelled, "Hey, catch!" Caught it right in my nose. So
every time I'd hear that word, I'd turn around. I guess
people decided to make it easier on me. Or harder.

                         CHRIS
Well, uh . . . nice to meet you, Catch.

                         CATCH
So what's a clean-cut little bugger like you doing all

the way out here? Shouldn't you be home sleeping in
your cozy little bed or something?

                         CHRIS
How do you know I don't live around here?

                         CATCH
Very funny.

                         CHRIS
I just felt like taking a walk.

                         CATCH
You felt like taking a walk <u>here?</u>

                         CHRIS
Yeah. I wanted to go somewhere quiet.

                         CATCH
Well, kid, this is definitely somewhere quiet alright.
But it ain't the peacefully-collect-your-thoughts kinda
quiet. It's more like the no-one-will-hear-you-scream
kinda quiet.
              (CATCH smiles at his little joke and no-
              tices CHRIS not smiling.)
Why did you want to go somewhere quiet?

                         CHRIS
I don't know.
              (Sees CATCH waiting for his real answer,
              then continues matter-of-factly.)
My dad died last week.
              (Pause.)
I mean, I didn't really know him all that well. I only
saw him once every few years or so. I guess I just
hadn't really had a chance to get out of the house and
be by myself for awhile.

                         CATCH
              (As CHRIS looks down.)
You're not gonna start crying are you, kid?

                         CHRIS
No. See, that's it. It's not so much that he's dead.
It's more the fact that I didn't really feel all that
bad when he died. It's just that . . . I guess I feel bad
for not feeling bad. I mean, a son should cry when his
father dies. It's like a natural law or something.

                         CATCH
Hell, I didn't cry when my father died. He was an
asshole. Tried to run my life.

                    CHRIS
I just don't know, you know.

                    CATCH
Hey, something brought you here didn't it, kid? Why
would you walk all the way out here for someone you
didn't care about, huh? It's like they always say, you
know, there's something good to find in everybody. I
know that probably sounds like a stupid old cliché to
you but, hey, it's true.
          (Notices the crack of a smile on CHRIS's
          face.)
Why do people hate clichés so much? I mean, if they
work, use 'em. I don't see why people get so pissed
off about someone who uses a line that, although it's
been used before, is probably the most easily under-
standable thing to say to someone. I mean, if it ain't
broke, don't fix it, right?

                    CHRIS
          (Laughing a little.)
You know, you have a very odd perception of the world.

                    CATCH
Yeah, well, somebody has to, kid. Listen, do you want
to sit down or something?

                    CHRIS
No. No, I probably should get back home before my mom
finds out I'm gone. I just hope she hasn't woken up and
checked my room or else I'm screwed.

                    CATCH
Yeah, I should probably go on duty pretty soon,
anyways. Gotta catch those suckers going in early for
work, you know.

                    CHRIS
Yeah.

                    CATCH
Gotta make a living somehow.

                    CHRIS
Well . . . bye.
          (Starts to go, but turns back around.)
Hey, Catch?

                    CATCH
Yeah?

<pre>
                    CHRIS
Was there anything good in your father?

                    CATCH
          (After a pause.)
Some people seemed to think so.

                    CHRIS
Hmm. Well . . . bye.

                    CATCH
Yeah, bye.
          (As CHRIS walks away.)
Hey, Chris.
          (CHRIS turns around.)
Nice talking to you.
          (CHRIS turns and EXITS.)

               THE END
</pre>

EVALUATION

This sensitive scene begins slowly, but even at the start, through the juxtaposition of the two dissimilar characters in a bleak environment, we get a sense that something is out of kilter. The very first action—a misstep of dropping money into the coffee—accentuates the feeling of unease. Catch, too, must sense that something is out of place, for he engages the young, quiet boy. Just as we begin to wonder where the scene is headed, Catch asks the question we are thinking: What are you doing here? In response, Chris reveals his feather's death, a potent fact that does what good drama should do; it both confirms and surprises. It confirms that something is wrong with Chris, and it surprises us with what that something is. In the end, with his talk of clichés, Catch reassures and helps Chris understand that his wandering is an expression of his feeling of loss and loneliness. In other words, his early-morning trek fulfills its purpose through his interaction with Catch.

Catch certainly provides us with an atypical bum. He's young, articulate, and caring—a sort of panhandler with a heart of gold. Does he profit from the transaction—other than the coin in his coffee? From the way he makes a determined effort to continue the conversation, he seems to enjoy and value the connection he makes with this troubled boy. Perhaps he sees himself in the boy. Although Catch's answer is cryptic, perhaps he achieves a slightly different perspective on his own relationship with his dead father as a result of this interaction.

The author uses relatively common language to good effect. Catch's name provokes interest, as does his calling Chris a "diaper bag." His admonition that this location is a "no-one-will-hear-you-scream kinda quiet" both entertains and augments the suspense. His lines about clichés are hardly profound, but they touch a proper chord in Chris, and Chris revisits the comments in his final question to Catch. Some of the lines contain relatively lengthy speeches, but they seem appropriate here, perhaps because of the quiet mood or perhaps because the two men don't know each other well enough for easy banter. Are Chris and Catch in conflict? Other than a little verbal sparring, not really, but the conjunction of these two lost souls, from the use of the prop cup of coffee to the clever language, achieves a touching effect.

Now let's look at a scene that sprang from a student's intimate knowledge of an environment and her careful observation of character.

EXAMPLE

THE DONUT SHOP
By Andrea Fisher

THE SETTING is a donut shop with an L-shaped counter center. Stage right are three one-piece plastic table and chair sets. Behind the counter are various machines and dispensers. The donuts are displayed in an alcove in the upstage wall. Small swivel stools are spaced regularly around the counter. There is a cash register on the upstage end of the counter. Behind the counter stage left is a doorway to the kitchen. The street entrance and exit is downstage right. The restaurant has the atmosphere of a truck stop, save that it is brighter and more plastic than most. The main colors are red and white. The restaurant light is bright and fluorescent. The time is 7 P.M. on a Friday night in early August.

(Darkness. BILL's voice comes out of the darkness and the lights fade up as he speaks.)

BILL

Well, I'm walking behind her, y'know, kind of bird-dogging her. About this time, she starts slowin' down. And then I start slowin' down, too, 'cause you know, this is a big woman, Harlan, and if she doesn't like somethin' you're doin', she doesn't hesitate to let you know. Yolanda may have been a lot of things, but she wasn't shy. So, anyway, here we are, her slowin' down, and me slowin' down, and I'm wonderin' if we're both gonna come to a dead standstill or what's gonna happen, when suddenly she turns around and looks at me kinda lazy and says, "Warm, ain't it?" Can you beat that?

Here I am, about ready to melt into a little puddle
right there in front of her, and she says, turnin'
around real slow, you know, with her eyes half closed,
"Warm, ain't it?" Whew, I'm tellin' you!

> (The lights reveal BILL and HARLAN at a
> table. BILL is a big man who tries to
> spread good cheer. He always has a story
> and jovially repeats himself if he has
> nothing else to say. HARLAN is short and
> muscular, with bright eyes. He is always
> filthy—he works at a garage—and he wears a
> cap with an oil company insignia. Much of
> his conversation consists of a series of
> cackles and whoops. He is a good-natured
> animal, and his wild laughter is particu-
> larly provoked by anything remotely con-
> nected with sex.

> ENTER EMMETT. EMMETT has the unmistakable
> look of a misfit. He carries himself with an
> open, slightly lunatic air of goodwill. He
> has a sense of drama, and is given to mys-
> terious pauses and knowing nods. He speaks
> loudly and smiles a lot. He is 30 years
> old, but seems younger. He carries a small
> bag in one hand.)

                    HARLAN
Hey, Emmett, when you gettin' married?

> (EMMETT grins, sits at the counter, takes a
> napkin from a holder and spreads it care-
> fully in front of him.)

                    BILL
Yeah, Emmett, when's the happy day?

                    EMMETT
> (Catching sight of a fly.)
Ah, there's that little booger now.
> (He stalks the fly, and after a series of
> unsuccessful thrusts with his hand, finally
> succeeds in catching it.)
Aha!
> (He takes it to the door and lets it go.)
                    BILL
You're really deadly there, Emmett.

                    HARLAN
Is that how you gonna catch you a wife?

                    EMMETT
          (Singing in a pleasant sotto voce as he
          fiddles with the salt and pepper.)
Gimme that old time religion,
Gimme that old time religion,
Gimme that old time religion,
It's good enough for me.
          (Calling.)
Hey, how do you get some service around here?

                    HARLAN
Yeah, Emmett, we know who you want service from.

          (He explodes into laughter as CEECEE ENTERS
          from the kitchen. She wears a white wait-
          ress uniform and a red apron. She is 16,
          very pretty, and a tireless and heavy-
          handed flirt.)

                    CEECEE
You guys talkin' 'bout me?

                    BILL
Well, maybe, Ceecee.

                    CEECEE
Now, Bill, you just watch what you're saying about me.

                    HARLAN
Ceecee, we was just talkin' 'bout what nice legs you
got.

                    CEECEE
Well, thank you, Harlan. Hiya, Emmett. Haven't seen you
for at least a coupla hours. Where ya been?

                    EMMETT
Oh, around.

                    CEECEE
Whatcha got in the bag, Emmett?

                    EMMETT
Oh . . . things.

                    BILL
What do you think of Emmett, Ceecee?

                    CEECEE
Oh, I think he's really somethin'.

                    BILL
We been tellin' him he'd make some girl a fine husband.

                    CEECEE
I bet he would, too. Who you gonna marry, Emmett?

                    EMMETT
          (Embarrassed.)
I ain't gettin' married.

                    BILL
Aw, come on, Emmett. A grown boy like you ought to
settle down. Now, when's it gonna be?

                    EMMETT
I tell ya, I ain't never been married, I ain't married
now, and I ain't never gonna be married, and that's all
there is to it.

          (As punctuation he yanks from his bag a
          doll that vaguely resembles CEECEE and sets
          it on the counter.)

                    BILL
Where'd you get the doll, Emmett?

                    EMMETT

Won it at the fair.

                    CEECEE
Oh, Emmett, isn't that cute. You brought me a doll!
How'd you know it was my birthday?

                    EMMETT
It ain't your birthday. It's my birthday.

                    CEECEE
Well, mine's next week. When's yours?

                    EMMETT
Today!

                    CEECEE
Today, huh? Well I just think it's awful mean not to
give me that doll. He won't give it to me, Bill, and
I've always wanted one of those dolls.
          (She grabs the doll.)
Why look. It looks just like me. I could be its mama.

                    EMMETT
          (Grabbing it back.)
I don't see how.

          (HARLAN starts to laugh.)
                    CEECEE
Oh shut up, Harlan.

(HARLAN attempts to restrain himself as
CEECEE pouts.)

Who you gonna give it to, Emmett?

EMMETT

I'm not giving it to nobody.

CEECEE

Well what good is a doll to you?

EMMETT

You never know. You never know.

CEECEE

You're just downright mean, Emmett.

EMMETT

I'm not mean. I just . . . well . . . what . . . what are you
giving me for my birthday? Nothing, I'll bet.

CEECEE

Oh, I'll think of something.

(HARLAN whoops and CEECEE snaps a towel at
him.)

Hush, Harlan.

BILL

Well while you're thinking, Ceecee, could you get me a
cup of coffee?

CEECEE

Okay. You want anything, Harlan?

HARLAN

I sure do!

(He whoops and starts laughing again.)

CEECEE

Oh shut up. I mean you want anything to drink?

HARLAN

Yeah, get me a Pepsi, I guess.

(CEECEE gets the drinks as the scene
continues.)

BILL

Well, Emmett, when're you gonna stop playin' with dolls
and get you some of the real thing?

EMMETT

Aw, come on, Bill.

CEECEE

Well, that's it! I know what I'll give Emmett for his

birthday! Emmett, I'm gonna give you one kiss for every year old you are, just as soon as I get off.

                    HARLAN

Hoo-wie! You heard it, Emmett. You gotta marry her now!

                    EMMETT
          (Overlapping HARLAN.)

Oh no you won't, Ceecee. You're not even gonna get near me. No sir!

                    BILL

Well, I call that one DEE-luxe birthday present, Emmett.

                    EMMETT
          (Rocking back and forth and giggling.)

Oh no. Oh-h no she won't.

                    CEECEE

Yes I will, too, Emmett, just as soon as I get off, so you better pucker up.
               (BILL and HARLAN laugh as EMMETT continues
               to rock and protest.)
I just don't know how I'm gonna control myself till then. I just don't know how. . . .

                    BILL

Why don't you just do it now, Ceecee?

                    CEECEE

Huh?

                    BILL

The boss ain't around. And I don't think anybody here's gonna object if you plant one on Emmett.

                    CEECEE

Well, I don't know if I should . . . you know . . . till I get off.

                    HARLAN

Come on, Ceecee. Give the boy his present.

                    CEECEE
          (Gets a comb and compact from her purse
          behind the counter.)
Well, let me make sure I look all right.

                    BILL

Hurry up, Ceecee, the poor boy's chafing at the bit.

                    CEECEE

Maybe I oughta go put on some rouge.

BILL

Whatsa matter, girl? Emmett too hot for you to handle?

HARLAN

   (Going behind EMMETT and putting his hands
    on his shoulders.)

He's all yours, girl. Come and get him!

EMMETT

Oh-h, no!

CEECEE

Well . . . What d'ya think, Bill, should I take my apron
off or leave it on?
   (HARLAN hoots.)

BILL

Well, I think you oughta at least loosen it a little.
You know, wear it kinda low.

CEECEE

Huh?

HARLAN

Come on, Ceecee, it's now or never.

CEECEE

Okay. Okay, Emmett, ready or not, here I come.

EMMETT

Oh-h, no!
   (He springs up.)

CEECEE

Oh, yes, Emmett. Here I come.

   (CEECEE comes slowly around the counter
   saying, "Here I come. Get ready, Emmett."
   Finally EMMETT lurches forward and grabs
   his doll.)

EMMETT

Oh-h, no! Oh-h, no!

   (He dives over the counter and runs out the
   door.)

HARLAN

That was some birthday present, Ceecee.

BILL

Well, what d'ya know? Just about every man between here

and Shanghai trying to get it, and Emmett turns it
down.
                    (HARLAN cackles, and CEECEE flings a dish
                    towel at him.)

                          THE END

EVALUATION

A major difference between this scene and the previous one is that
The Donut Shop has four characters, and *Cliché* has only two. As you know
from previous work, relationships are either direct or modulated. Chris and
Catch have a direct conversation. What Chris says is meant for Catch, and
what Catch says is meant for Chris. Even when a line such as "I just felt
like taking a walk" has underlying meaning, both the surface and the
underlying meanings are exchanged between just those two characters.

As the number of characters increases, the structural possibilities
increase geometrically, as you saw in Chapter 5. Bill has a relationship with
Harlan, and Harlan has one with Bill. Each of those relationships is altered
by Emmett's entrance. In addition, Emmett has a relationship with each
man, and each man with him.

All of that becomes even more complex with Ceecee's appearance.
She has a relationship with each of the three other characters, and each of
them has a relationship with her. All of those relationships are affected by
the presence of others in the room. Without Bill and Harlan around, for
example, Ceecee might tease Emmett a bit or she might ignore him, but she
would hardly approach him as if to kiss him. She does that because Bill, by
calling her bluff *in front of others* dares her to it, and she cannot back down
without losing credit. Her performance is at least as much for Bill and
Harlan's benefit as for Emmett's.

This scene lives by its characters. The author has observed carefully
and has selected telling details about each of them. There's Emmett, a
grown man, but childlike in his simplicity. Like an adolescent, he is fasci-
nated by Ceecee and yet scared of her. His common language such as
"Oh . . . things," and "You never know" produces a mysterious imprecision.

Ceecee is at that stage where a blossoming young girl tests her
effects on men. She's flirtatious. She wants people to think she knows
everything about love and life. She wants the attention of men, but she also
wants to control the relationship. Bill is a relatively normal individual who
provides links between the characters. He appears to have a good instinct
for human nature, and he occasionally turns a clever phrase.

Harlan seems sharper mentally than Emmett, but emotionally he is
hardly more mature. His cackles and whoops sharply define his character.
Collectively Bill and Harlan provide an audience before whom Ceecee and
Emmett can act out their drama.

The author could have written this scene with three characters,
omitting either Bill or Harlan, but I think both are necessary. Harlan gives

us the crudest and most basic possible vision of relationships—everything is based on sex. Bill has a more sophisticated sense of human relationships, which is why he is the one who calls Ceecee's bluff.

But characters, by themselves, do not make a scene. They must be spun into action, into conflict. In that respect the writer has developed these individuals carefully and cleverly. Emmett's singing and fly catching instantly mark him as an eccentric. As in some other scenes, questions provide an opening hook. Although they are asked in jest, the questions put to Emmett about marriage and a wife—and his ignoring those questions—establish character interactions and focus the scene on male-female relationships.

The author does not try to introduce all the characters at once. We first meet Bill and Harlan. Then Emmett. Then Ceecee. Such a progression allows the audience to get to know the characters more easily than if all four were introduced together in a group. Virtually as soon as all the characters are identified and Emmett has answered the opening question ("I ain't never gonna be married!"), the doll is brought out. Like the candy hearts, the cookies, and the sandwich in previous scenes, the doll inaugurates the conflict and becomes a concrete symbol of it. Ceecee wants it not only to tease Emmett but also to prove to herself and to Bill and Harlan that she controls the relationship.

When Emmett asks her for a present, Ceecee comes up with the kiss idea, knowing it will both tantalize and embarrass Emmett. She also thinks she will never have to follow through on it.

A key moment of the scene occurs when Bill says, "Why don't you just do it?" Ceecee is caught. She backpeddles and delays, but she finally goes ahead. She doesn't want to lose face with Bill and Harlan. She also sees that the thought of a kiss from her is unsettling Emmett, and Ceecee likes to unsettle men. The playwright has remembered the first lesson; she has put the characters into action. The picture of Ceecee torturing Emmett, slowly approaching with her lips puckered as Bill and Harlan attempt to hold him, has solid theatrical value.

Having built to the climactic moment of the kiss, the author deflates the bubble by having Emmett bolt. In a way, all the characters in this scene win. Ceecee confirms her power over men and retains her standing with Bill and Harlan. The two men achieve a good laugh and teach a small lesson to Ceecee in the process. Emmett avoids contact with a girl, but at the same time he has the joy of thinking that Ceecee was really going to kiss him.

No final resolution is achieved. The scene leaves us wondering whether Emmett will ever get his kiss or if he'll somehow get revenge on Ceecee for taunting him. If a scene leaves you wanting to know more, as I think *The Donut Shop* does, that's a good sign. And that's when scenes begin to grow and expand into plays.

7

. . . .

Writing from a Source

EXERCISE 7.1 • WRITING FROM A SOURCE
Write a scene using an item from a newspaper or a magazine as
the springboard. Use as many characters and as much dialogue
as you need. Minimum length: 5–6 pages. For comparison
purposes, keep a copy of the news item.

In this chapter your assignment is to write a scene using an item
from a newspaper or magazine as a starting place. An historical incident
would also work. It may seem that we're heading in the wrong direction.
After all, in Chapter 6 you learned that everything you write must come
from your own experience, from your own thoughts and feelings.

That's still true. But there are different ways to get at your thoughts
and feelings, and the purpose of this chapter is to work in the opposite
direction from the preceding exercise. In the exercise in Chapter 6, the
impetus is internal. The germinal idea is something you're very familiar
with, and, in the course of developing that idea into a dramatic piece, you
must provide it an external reality that other people can recognize.

In this exercise the seed is something *external*—a news item, for
example—and you must make it *internal*. You must create environments,
situations, conflicts, and characters with depth and reality. Perhaps giving
"reality" to something that begins with "reality" sounds redundant. Or at
least simple. Yet you will be amazed how easy it is to begin with a "real"
incident and wind up with an "unreal" play.

Authors writing about George Washington or Wolfgang Amadeus
Mozart cannot merely relate incidents. They must create from their own
experience—which in these cases surely includes readings about
Washington or Mozart—a Washington or a Mozart that is an understand-
able human being to an audience. That, of course, is what Maxwell Anderson tried to do in *Valley Forge* and what Peter Shaffer tried to do in
Amadeus.

In one of my classes a student tried to write a scene based on a
Middle Eastern terrorist. The student attempted to develop the human side
of the young activist. The scene broke down because the young writer was
unable to breathe "reality" into this "real" person. The writer had little
sense of religious fervor, of political calculation, or of group honor, all of

which propelled the terrorist to action. Research into Mideast politics and the psychology of terror might have helped the writer provide a specific and dramatically viable focus to the scene.

Another student wrote a scene about a wealthy businessman who had committed a crime. But his dialogue about "big deals" and "lots of money" was so general that the "real" businessman immediately lost all credibility and seemed fake.

In other words, if you write a scene about a terrorist or a businessman or even George Washington, you must come to know that person like a good friend or a next-door neighbor so that you can show us how that person thinks, talks, and acts.

Examples of plays that proceed from real events to dramatized events are legion. For *Amadeus,* Peter Shaffer used as his seed the old musician's rumor that Wolfgang Amadeus Mozart had been murdered. As the play developed in Shaffer's mind, the apparently subsidiary character of another court musician, Salieri, became the linchpin of the piece.

David Henry Hwang's sensational *M. Butterfly* was prompted by a two paragraph story in *The New York Times* about a French diplomat who had fallen in love with a Chinese "actress" who turned out to be both a spy and a male. But Hwang didn't just dramatize an incident. Rather, he devised a situation that commented on truth and illusion as well as cultural perceptions and misperceptions between Asians and Europeans. Similarly, Timberlake Wertenbaker's *Our Country's Good* latched onto an account of Australian convicts in 1789 giving a performance of George Farquhar's 1706 comedy, *The Recruiting Officer.* In constructing her moving play-within-a-play, Wertenbaker researched the early 18th-century London of Farquhar and the late 18th-century London that spawned the convicts as well as Australia's penal conditions.

Anna Deavere Smith interviewed people involved in traumatic social upheavals for her plays based on riots in Brooklyn and Los Angeles. The productions of her work create a kind of docu-drama, which combines the direct contact of live theater with the rapid movement usually associated with film and television. In *The Crucible,* Arthur Miller used news items and historical material in two ways. Enraged by the Communist-hunting tactics of the House of Representatives Committee on Un-American Activities in the late 1940s and early 1950s, he wrote *The Crucible.* The play dealt with the same topics that Miller saw treated everyday in the newspaper—persecution, guilt by association, and a kind of mass hysteria. But instead of setting his play in the 1950s in Washington, D. C., he set it in 1692 in Salem, Massachusetts, during the days of the Puritan witch hunts.

DANGERS OF FACT

Just as there is a danger, when working with material that comes from your own experience, in hewing too much to actual fact, so there is a danger in working with external materials in trying to be *too* factual. You *can* change history. Shakespeare did it all the time. Jerome Lawrence and Robert E. Lee wrote *Inherit the Wind* based on the famous Scopes trial of evolution and creationism. Whenever they worked with historical material, Lawrence once said, they did as much research as possible—and then threw it out to write the play. For *Inherit the Wind,* for instance, they read the entire court transcript. But in all that riveting courtroom drama, they only used four lines from the actual transcript.

There are, however, some points you might consider. Generally, the closer in time and the more prominent the incident, the less leeway an author has. If you were writing about the conflagration in Waco or the Oklahoma City bombing, you could hardly change the names of the defendants or other principle participants, and the incidents you incorporated would need to parallel actual events pretty closely. Even at that, you would certainly be inventing dialogue and action based on your understanding of the people and events.

If you were writing about the competition sparked by the beginning of human flight, you'd have to use Orville and Wilbur Wright, but the various people around them are virtually unknown now. The distance in time allows greater dramatic license. You could probably invent or combine characters or even create or alter incidents, depending on your dramatic purpose.

I have a final warning regarding historical materials. Frequently when dealing with historical personages, writers lose their sense of language and dialogue. Suddenly everyone is expressing grand thoughts in complete sentences with no contractions. Yes, dialogue must reflect the time period of a play, but it must first give us entry to an apparently real human being. Even if you immerse yourself in the language of a particular time period by reading newspapers, letters, and diaries, remember that such written words are always more carefully constructed than casual, spoken dialogue.

Often when I assign this exercise a student will ask me about doing an adaptation of some other work, such as a scene from a novel or a short story. Adapting a story from one medium to another makes a valid and interesting exercise, but it is a very different exercise from this one. When you adapt something, the story itself is complete. The challenge is to move that story from one medium to another. It's as if you had a complete jig-saw puzzle, and you were trying to re-do the picture as a sculpture or an oil painting. In this exercise the story is *not* complete. You must create at least part of it. Think of this exercise as having one piece of a jig-saw puzzle and being asked to design a picture around that one piece.

At the start of this chapter, I said that working from a source amounted to taking something external and making it internal. In fact it's really a three-part process. First, you find something external, such as a news article, that piques your interest. Second, you make it internal by imbuing it with knowledge, with creative imagination, and with feeling. Third, you translate it into a scene or a play, thereby making it once again into an external product.

Now let's observe that process at work in two student-written pieces.

• • • **EXERCISE 7.1**

Write a scene using an item from a newspaper or a magazine as the springboard. Use as many characters and as much dialogue as you need. Minimum length: 5–6 pages. For comparison purposes, keep a copy of the news item.

The scene should not attempt simply to reconstruct what happened. Rather, allow your imagination to play with the account until you find something dramatically interesting to work with.

EXAMPLE

This piece sprang from a news item that children in Palermo, Italy, had hired a witch doctor to cast a spell on a teacher they hated.

ABRACAPOCUS

By Bill Roundy

THE SETTING is a grimy restaurant. Three teen-age boys sit in a corner booth, each with a mug of coffee. Remnants of food litter the table. NATHAN wears a large black leather jacket with small furry objects hanging from it. BROOK and ROB wear more ordinary clothes. BROOK is looking at a paperback book. At a nearby table a WOMAN sits reading a newspaper and eating.

BROOK

Would you look at this? It's impossible. I can't even . . . It's impossible.

ROB

What is?

BROOK

This book!

     (Tosses it on the table.)

Look at that. It's gotta be, like two hundred pages. And I've got to write ten pages on it by Wednesday.

ROB

Have you even started it?

     (BROOK snorts.)

NATHAN

That means "of course not."

BROOK

I mean, here we were, going along with nice, boring Emerson and learning about his stupid transcendental

philosophical crap—nice, easy, snooze material, and
then WHAMMO! Out of the blue we get hit with this mon-
strosity. Ten pages of literary analysis on a book I
haven't even looked at. It's impossible. I can't do it.
I'm not even gonna try.

                    ROB
What's the book?

                    BROOK
The Scarlet Letter.

                    ROB
Oooh! Not good! Not good at all! It blows dog.

                    BROOK
Have you read it?

                    ROB
Back at my old school.

                    BROOK
Say, could you . . .

                    ROB
I am not writing your paper for you.

          (BROOK growls and takes a sip of his
          coffee.)

                    NATHAN
When's it due?

                    BROOK
Day after tomorrow.

                    ROB
You could skip school.

                    NATHAN
No he can't.

                    BROOK
Why not? That's a great idea.

                    NATHAN
Two reasons. First, it would just be due again the next
day. Second . . .

                    ROB
WAAUUGH! COCKROACH!

          (ROB points to something scuttling out of
          his sandwich. The WOMAN turns to look at
          ROB as NATHAN quickly upends his coffee cup
          and slams it over the roach, trapping it.)

                              NATHAN
And second, well, just think about the last time.

                               ROB
Guys, there was a cockroach in my food!

                              NATHAN
It's okay. I caught him. Anyway . . .

                              BROOK
Which last time?

                              NATHAN
Remember? The last time you skipped school?

                              BROOK
          (Slamming down a fork as he recalls a trau-
          matic experience.)
Oh, no! Yeah, I remember. Forget that!

                               ROB
Why, what happened?

                              BROOK
I don't want to talk about it.

                              NATHAN
He wanted to catch the opening matinee of <u>Lost World</u>,
so he . . .

                              BROOK
I don't want to talk about it.

                               ROB
Why?

                              BROOK
BECAUSE I DON'T WANT TO TALK ABOUT IT!

               (The WOMAN glances over at him.)

                              NATHAN
There's another solution.

                              BROOK
Short of actually reading the book, you mean?

                              NATHAN
Yes.

                              BROOK
What?

                              NATHAN
          (He looks around suspiciously and lowers
          his voice.)

What if Mr. Rosenthal didn't come in on Wednesday? Then
you couldn't turn in your paper, could you?

                         BROOK
Well . . .

                         NATHAN
And if he were, say, sick on Thursday and Friday, too,
than you wouldn't have to worry until next week. Which
would give you plenty of time to find some Cliff's notes
and rip off a decent paper.

                          ROB
What do you want him to do, get hit in the knees with
a lead pipe?

                         NATHAN
Oh, nothing so crude.
                  (He reaches into an inside pocket of his
                  jacket and removes a small black book,
                  which he drops in the middle of the
                  table.)

                         BROOK
Cliff's notes with a funny cover?

                         NATHAN
The Lesser Key of Solomon. Produced by Golden Goblin
Press, 1996, based on the works of Aleister Crowley
combined with ancient Middle Eastern lore. This is the
only copy I've ever seen.

                          ROB
Where did you get it?

                         NATHAN
                  (Smiles and shakes his head.)
I can't talk about it.

                         BROOK
So, what are you gonna do with that?

                         NATHAN
Wrong question. It's what are you gonna do. In that
book there's a spell that can give the target about
three days' worth of some really nasty cramps. It
should take care of Rosenthal, no problem. Give him
diarrhea, too.
                  (NATHAN stands.)
It's yours to use, you pick up my part of the bill.
And I want the book back tomorrow.
                  (He shuffles out of the booth and EXITS.)

                              BROOK
Whoa.

                               ROB
That's weird. What's in it?

                              BROOK
               (Carefully opens the book and flips through
               some pages.)
Let's see. Maybe there's an index. Nope. Hmm.

                               ROB
C'mon, just read it.

                              BROOK
Wait. Here: "To Cast an Incantation of Deadly Illness."
I don't know. I mean, I don't want to kill him.

                               ROB
As if it would really work anyhow.

                              BROOK
It might. You never know. Wait a minute.

                               ROB
What?

                              BROOK
It says here we need a goat.

                               ROB
A goat?

                              BROOK
Yup. "Take thou a goat without blemish or deformity . . .

                               ROB
I've got a hamster at home.

                              BROOK
. . . and slay it with a sacrificial knife."

                               ROB
EEEWWW! No! You can't have my hamster.

                              BROOK
And then we're gonna need mandrake root and pig fat
and . . . something else. . . . What? "Brew for six hours?"
I could read the damn book in that time!

                               ROB
Okay, so we make some substitutions. We got this cockroach here. What else do we need? Mandrake? That's like
a spice, right?

                              BROOK

I guess so.

                              ROB

Cool, no problem. We just use pepper, and um . . . yeah,
there's some bacon strips from my sandwich—that's pig
fat.

                              BROOK

Whoa. You think it'll work.

                              ROB

No.

                              BROOK

Oh.

                              ROB

But what's the hurt in trying, huh?

                              BROOK

All right.
          (Looking at book.)
Um, "sacrifice thou the . . . uh . . . cockroach."
          (ROB lifts the cup, clutches at the wrig-
          gling form, lifts his fork and impales the
          insect.)
Cool! Now, throw on the other stuff and I'll chant.
"Adonai, Moroni, forenchin senotas,
malefactarium . . . uh . . . Joe Rosenthal . . . uh . . . bardide
antedevarian petrovich."

          (As he chants, BROOK makes strange motions
          with his freehand over the center of the
          table while ROB breaks up bacon pieces and
          dumps pepper on the roach.)

                              ROB

That's it?

                              BROOK

Yup.

                              ROB

How do you know if it worked?

                              BROOK

Well, I guess we . . .

          (Suddenly the WOMAN at the next table
          pitches onto the floor clutching at her
          throat and twitching violently. BROOK and
          ROB look at each other and then at the
          WOMAN lying prostrate on the floor. Both

```
                boys get up, throw the books in a bag, and
                begin to move offstage.)
                              ROB
        Check, please!


                            THE END
```

EVALUATION

There are several remarkable aspects to this little scene. First, the author's basic decision to move the scene from Italy allowed him to work with characters with whom he was familiar. The three main characters are nicely differentiated with the mysterious Nathan, the uptight Brook, and the normative figure of Rob. The obstacle is Brook's assignment, a problem that is easy to relate to because all of us face deadlines of one sort or another. The author also places a time pressure on the situation, a reason for this action to occur right now. Brook's paper is due in two days, which provokes his anxiety into action. There's little direct conflict among the three boys. For the most part they are cooperating in trying to solve Brook's problem. The dialogue flows naturally as, for instance, when Brook begins a sentence ("Say, could you . . .") and Rob responds even before the thought is complete ("I'm not writing your paper for you.").

The scene progresses in a rather uneventful mode as the boys discuss the problem amid apparently tangential distractions such as the cockroach and the woman at the next table. The mood shifts, however, when Nathan lowers his voice and produces his magic book. His description of it and the stilted quotations from the book provide an authentic ring even though the actual directions and the boys' reactions to them are comedic. Their substitutions are suitably clever, and they bring the cockroach nicely back to a central role in the scene. We, like the boys, expect absolutely nothing from the curse, so the unanticipated writhing of the woman surprises us, but we also realize we have been prepared for it all along.

The scene contains several nice details. Brook's book just happens to be *The Scarlet Letter*, which also deals with an unseen curse. Then there's Brook's mysterious skipping of school, which establishes him as a person who might do something crazy even though we never find out exactly what happened.

EXAMPLE

This scene began with a student's interest in an article about dieting, the kind found in almost any supermarket magazine.

```
                      UNATTAINABLE
                    By Wendy Brodsky

        THE SETTING is a split stage. Each side
        shows a bedroom of a teenaged female with a
```

bed, a phone, and a full-length mirror. On one side, LIZ sits on her bed with an open magazine in her lap and a phone to her ear. She is a thin, pretty girl. Her eyes are focused on the mirror as she examines rolls of skin produced by her sitting cross-legged. In the other room, BECKY lies on the floor holding the phone to her ear with her shoulder. She is similar to LIZ, but a little heavier. She is struggling with her legs in the air in a contorted position. After a moment BECKY lets out a moan and readjusts the phone.

                    BECKY
Uh, Liz, are you sure this is how they say to do it, 'cause, uh, it's really pretty painful and I can't imagine they think my body would do this on purpose.

                    LIZ
          (Glances from the mirror to the magazine.)
Yeah, I'm sure. And it's not that painful. Let's just hold it the full five minutes and see how we feel. It's supposed to be the best toning exercise around. Just think about how good we'll look.

                    BECKY
Yeah, and how much we're gonna hurt tomorrow morning when we wake up. If we wake up.
          (BECKY's legs start to drop. Her face is
          clenched as she continues the exercise.)
I can't be doing this right. I'm stopping. I don't care how toned its gonna make my butt.
          (BECKY pulls her legs to her chest, break-
          ing the exercise.)

                    LIZ
Wimp. You only had a few more seconds. I guess it's just not as important to you as it is to me.

                    BECKY
Guess not.

                    LIZ
          (Still seated on the bed, LIZ looks again
          at the magazine and pretends to be out of
          breath.)
Fifty-five, fifty-six, fifty-seven, fifty-eight, fifty-nine, five minutes. Ah, that feels good. Okay, are you ready to try the next one?

                    BECKY
          (Lying on the floor.)
Naw, I don't think so.

                    LIZ
How come?

                    BECKY
I just don't want to, okay? It hurts.

                    LIZ
Whatever. But I'm gonna try the next exercise.

                    BECKY
Go ahead. Do you want me to talk to you while you do
it?

                    LIZ
Actually, it's pretty distracting. Maybe you should
just go.

                    BECKY
I probably should anyway. My mom's been calling me to
dinner for, like, the last half hour.

                    LIZ
I thought you swore off eating until after the dance.

                    BECKY
          (BECKY gets up. It requires more energy
          than she expected, and she immediately sits
          on the bed.)
Yeah, I did, but I've got to at least pretend or
she'll never leave me alone.

                    LIZ
Yeah, I know what you mean. My mom's the same way.
          (LIZ walks to the mirror. She runs her hand
          across her face and down over her stomach
          and thighs.)
So . . . I'll talk to you later.

                    BECKY
Yeah, sure. Are you okay? You seem, I don't know, upset
or something.

                    LIZ
I'm fine. I just . . .
          (Turning and examining her profile in the
          mirror.)
Do you ever wonder what your body looks like to other
people? I mean, like, how much cellulite they see that
you never even noticed before?

>   DECKY
> That's why I try to avoid mirrors and nakedness at the
> same time. You gonna do the next exercise?
>
>   LIZ
> Yeah, in a minute. But I don't know why I even bother.
> It's not like they help anything.
>
>   BECKY
> That's precisely why I quit. I need faster results.
> Liz?
>   (LIZ frowns and steps back from the
>   mirror.)
> Liz?
>
>
>   LIZ
> Huh? Yeah.
>
>   BECKY
> You sure you're all right?
>
>   LIZ
> I'm fine, Becky. Go eat your dinner.
>
>   BECKY
> I told you, I'm not eating. I'm just, you know, going
> down there.
>
>   LIZ
>   (LIZ lifts her shirt and examines her stom-
>   ach. She grabs whatever skin she can and
>   pinches it in her hand. She turns away in
>   disgust and looks to the magazine on her
>   bed.)
> Yeah. Me neither.
>
>   BECKY
>   (BECKY moves to the door while LIZ re-
>   turns to her bed. BECKY moves with normal
>   energy while LIZ drags, her body hunched
>   over.)
> But, like, I don't want my mom to get mad.
>
>   LIZ
> Oh, no. Definitely don't want that.
>
>   BECKY
> So, uh, I guess I'll go downstairs, just, you know, to
> put in an appearance.
>   (LIZ picks up the magazine and tears out
>   a handful of pages, crunches them into
>   little balls and throws them at the

> mirror. BECKY hears the noise through the phone.)

What are you doing?

                    LIZ

I was just, uh, tearing out the exercises so I could see them better, all at once, instead of having to, like, turn the pages.

                 BECKY

Oh. You still gonna do some more?

                    LIZ

Of course. Why wouldn't I?

                 BECKY

I dunno. I just thought maybe . . . Well, I'm going now.

                    LIZ

Okay. I'll talk to you later.

                 BECKY

Yeah, see ya.

> (Both girls hang up their phones. BECKY carefully places hers, but LIZ simply drops hers to the ground. On the way out of her room, BECKY pauses briefly to glance at herself in the mirror. Then, with a shrug and little laugh, she EXITS. Meanwhile, LIZ looks at the scattered magazine pages. She sits on the floor, picks up a page, and unravels it. She stares at it, then tears the page in half. She lets the pieces drop and then buries her head in her hands just as we see BECKY's door close.)

               THE END

EVALUATION

Normally I don't encourage the use of telephones in scenes. Too often they become merely a crutch to convey exposition. In this scene, however, the telephone plays a pivotal role because it allows Liz to masquerade as someone else. There are, in essence, *three* characters in the scene: Becky, the Liz that Liz wants Becky to see, and the actual Liz that the audience sees. The girls converse about whether they will or will not do the exercises, but the real conflict is an internal one within Liz. The author has captured the mood of a character in despair. Where Becky seems to realize that she will never be supermodel slim, she does the best she can and seems willing to live with the results. Liz, on the other hand, is conflicted to a point that leaves her virtually unable to act. She compares the images in the magazine with the image of herself in the mirror. Although she is actually thin already, in her mind the gap is so great that she has no hope of narrowing it, so she refuses even to try. That's bad enough. But even worse is her lack of honesty about her despair. She does not want to be perceived as "not trying," so she lies even to her friend. The talk of food and dinner, combined with Liz's deceptive behavior, suggests without saying it directly that Liz's problems almost certainly also include eating disorders. The final tableau, with Becky exiting to the dinner table while Liz isolates herself in her room, presents a visual contrast that represents the disparity of the worlds of these two young women. To her credit, the author was able to transform one of the hundreds of mundane articles on diets and dieting into a rich and personal examination of a human being at risk.

You've now written a range of pieces of varying complexity. The next chapter will give you several more writing challenges to probe your creative abilities and stretch your playwriting skills. After that, you'll be taking your scenes and molding and extending them into plays.

8

Expanding Your Skills

Before we move on to writing your play, I want to share with you several additional writing exercises. Some of these cut right to the heart of what dramatic writing is all about. Others are designed just to get the writing juices flowing. Some of them are exercises I've devised, while others I've gleaned from playwrights and playwriting instructors from around the country. All of them have been proven effective in developing playwriting skills. I've included some short examples where I thought they would be especially beneficial in illustrating the exercise, but most of the instructions are explicit without examples. The exercises are included in no particular order, so you can pick and choose which ones appeal to you. I think, however, you will find all of them helpful and worthwhile.

SPONTANEOUS COMPOSITION
The first exercise is designed to explore concerns that are important to you. It is also an excellent device to get you started writing when you don't particularly feel the inspiration. The credit for devising this exercise goes to Jon Sedlak, a playwright who introduced it to me when he visited my playwriting class.

• • • EXERCISE 8.1
Write a monologue, which is a relatively lengthy speech by one person. You should not consider meaning, structure, grammar, punctuation, or even character. Rather, simply sit down and start writing about your own thoughts and feelings. The finished product should be about a paragraph or page long. Repeat the process two or three times, thus generating two or three different monologues.

This exercise, by itself, is not designed to produce a play. It can, however, accomplish several objectives. First, it will help you overcome the hesitation of starting something. It will allow you to write without the pressure that what you write has to be "good." These monologues are not finished products but rather stepping stones. Although they are not designed for formal production or evaluation, you should look at them carefully to see if they suggest potential for drama and character.

That potential leads to a second objective of the exercise. In many instances you will generate a monologue that can be used in the development of a play. Perhaps a conflict, a theme, or an emotion will materialize. Perhaps a certain phrase or group of words will hint at dialogue or character.

Yet a third reason for Spontaneous Composition is that if you are unhampered by the conventional constraints of character, plot, and dialogue, you will often discover a depth of feeling previously unknown about a subject. You may find an interest of which you were completely unaware. In short, this exercise helps you become alert to your own subconscious feelings and concerns.

EXAMPLE

GET A LIFE
By Shannon Dickinson

Why is she so obsessed with my every move? Why does she care? I don't go around making comments on every decision she makes. She needs to worry about her own life and her own problems. I am an adult now. I make adult mistakes. I don't need anyone else pointing these things out. I need to learn from my own mistakes. Is that not life? Look at what she's done. Plenty of mistakes there. Did I judge or comment snidely then? I don't think so. I have done nothing to deserve this. I've always been there for her, never asked for anything in return. Total unconditional love. Isn't that what sisters are for? She needs to have a kid so she'll leave me alone. Someone she can control. I'm tired of filling that void. Three words for her: Get a life!

It's easy to see from that example the strength of feeling generated through this assignment. It's also easy to see how parts of the monologue could spark dramatic writing. The conflict, the characters, even situations begin to emerge. One sibling makes some "adult mistake," and her sister takes her to task for it. At some point the two of them confront each other utilizing the emotions and ideas of the monologue, which could easily be adapted into dialogue. The seeds of the drama are exposed right in the center of this short piece of writing.

PROBLEM SOLVING: PEOPLE, PLACES, AND PROPS

As you develop your play, you may find yourself running into some very mundane problems. "I'd really like to have a scene where John finds Barbara deceiving him with another man, but why would they get together if they've just had a fight?" If you need two characters in the same room at the same time for their climactic confrontation, you must solve the problem of how to get them there.

Playwrights must be practical problem solvers. They must be able to put all the pieces of the puzzle together. And just as more pieces make a jigsaw puzzle more difficult, a longer play makes the playwright's task more challenging.

This problem-solving facet of the playwright's job is a much underrated talent. Every time characters enter or exit, they need a reason for their coming or going—a plausible reason. A beginning playwright might explain that a character visiting a friend "just drops by." Even if friends do that in real life, in plays they need at least a semblance of a reason. Maybe the visitor is lonely and wants attention. Perhaps the visitor is delivering a package or is just plain bored and wants the other character to go somewhere or do something to help pass time. Even that is a reason.

I once examined Robert Bolt's well-crafted play *A Man for All Seasons*, which dramatizes Sir Thomas More's conflict with King Henry VIII over the king's marriage to Anne Boleyn. At one point (Act II, scene 7), More is in jail and his wife, daughter, and son-in-law visit him. They are permitted to see him only so that his daughter Margaret can try to persuade More to sign an oath of allegiance to the king.

The three are ushered into the cell area by a jailor, who lets More out of his cell for the visit. Greetings are exchanged, and the family give More some small gifts they've brought. Then we get to the heart of the scene. Margaret tries to persuade More to sign the oath. The real point of their good-natured but emotionally charged argument is to display a loving relationship between the father and the daughter. She understands why he cannot sign. Even as she urges him to do it, she realizes that she would think less of him if he did.

The playwright exerts pressure on the scene. The jailor announces that the visit can last only two more minutes. More sends his son-in-law to occupy the jailor, and then he tells his wife and daughter to leave the country. He persists until they accede.

More's wife, Alice, is furious with her husband. She doesn't understand why his conscience is more important to him than his family. More tries to placate her, but her anger will not be assuaged. Finally, frustrated by his inability to make his wife understand, More breaks down. At that point Alice rushes to him and offers her support, not because she suddenly understands, but because she realizes her husband needs her.

The main transactions of the scene are done. The relationships between More and his daughter and between More and his wife are finalized, so the scene can end. But Bolt does not end it gently. Disputes erupt as the jailor forces the family to leave.

There are 118 speeches in this scene. Fifty-three speeches, which is 45 percent of the scene, are devoted to the entrance, the greeting, the gifts, the "two-minute" warning, and the exits. And all of those items are purely mechanical elements constructed by Bolt to surround and support the two crucial human relationships between More and his daughter and More and his wife, which constitute the core of the scene.

That is not unusual. In fact, Bolt probably invests his scenes with more substance than most playwrights. Good playwrights like Bolt solve the technical problems, and they are willing to take time to do it. They put the pieces together seamlessly and construct effective transitions to move from one segment to another. Good playwrights even use the mechanical elements to enhance the effectiveness of the core elements. Bolt doesn't waste 45 percent of the scene. He uses the conflicts with the jailor, for example, to heighten the tension of the scene and turn a simple exit into an emotion-wrenching departure.

The following exercise is designed to test your problem-solving abilities. It is structured to be used with a group of people, in a class or at a workshop, but I have included instructions to allow individuals to do the exercise.

• • • **EXERCISE 8.2**

Write a description of two characters. Each description should be only a sentence long. Write one sentence each describing a place and an object. There does not need to be a connection between any of the four items described. The descriptions will be used to develop a scene. Bring your descriptions to class and exchange them with another student. Write a scene of about three or four pages using the characters, place, and prop you have been given.

This exercise is in many important ways a reverse of the Spontaneous Composition assignment. Where that exercise sought to probe impulse and depth of feeling outside the conventional restraints, this exercise is pure problem solving within the confines of a given location, given characters, and a given prop.

Within those defined limitations, the assignment is as much an exercise in creativity as the previous one. The student must solve the basic problem—What are those two characters doing in this place with that thing?—in a plausible, dramatic fashion.

Another strategic benefit of this exercise is that it increases a writer's awareness of environment and props. As soon as characters are placed within a locale, their behavior and therefore their actions are affected. And actions often involve doing something with objects. From the very first silent scenes in this book remember the woman with the Christmas tree and the ornaments?—to famous plays such as Laura with her glass menagerie, props help to focus action. They play a vital and much underrated role in good drama.

Writers often go in one direction or another with this exercise. Some just throw up their hands and settle for anything, but others create truly inventive solutions to the problem. That, of course, is the point: to address those problem-solving skills that every play requires.

Obviously this exercise is designed for a group rather than for an individual. However, to assist you in using it for yourself, I've included six sets of character descriptions, places, and objects with which you can create scenes to test your own problem-solving talents.

Characters	Object	Place
An insurance representative A man with a clerical collar	Wrapping paper and ribbon	A garage
A member of the armed forces A pregnant woman	A camera	A park
A schoolteacher A man in work clothes who performs odd jobs	Flowers	A yard sale
A well-dressed female business executive A souvenir seller	A piece of underwear	The deck of a swimming pool
A 60-year-old man recovering from an operation A clever 10-year-old	Room deodorizer	A funeral parlor
A hair stylist A woman not native to this country	A toy boat	A bar when it's closed

LOCATION, LOCATION, LOCATION!

The old real estate joke asked: What are the three most important assets in selling a house? Location, location, location! The same could frequently be said of plays. The *place* often serves to define the *action*. Many plays are situated in quite common environments—a house, a restaurant, or an apartment. Others are more specific. David Mamet use places of business to delineate his characters and situations, from a junk shop (*American Buffalo*) to a real estate office (*Glengarry Glen Ross*), to a professor's office (*Oleanna*), to a producer's office (*Speed the Plow*). Eric Bogosian sets his "slacker" drama *subUrbia* outside a convenience store. The action of David Edgar's arresting *Pentecost* takes place in a deteriorating church with a remarkable fresco in the middle of revolution. This exercise asks you to explore the importance of location in a scene.

• • • **EXERCISE 8.3**
Find an interesting location and develop a scene within that place.

The site you select need not be exotic. A child's bedroom can be just as dramatically powerful as a dank storm sewer. Look for a place that speaks to you, that ignites your creativity.

WORDS, WORDS, WORDS

"What do you read, my lord?" Polonius asks Hamlet, to which the Prince replies, "Words, words, words." In a very real sense, "words" are what playwrights are all about, and this next exercise focuses your attention on the verbal aspects of playwriting.

• • • EXERCISE 8.4

Look at something that has words on it, such as a pamphlet, a catalogue, or a brochure. Pick out two or three words that pique your interest or curiosity. These will become the basis for a scene. Work one of the words into an initial line of dialogue and enlarge the scene from there. Work another of the selected words into the dialogue within 8–10 lines.

Not only is this an excellent way to get the writing muscles warmed up, but it also forces you to think about and value words. Right now, for example, I'm looking at a list of classic movies, and here are some word combinations that would serve as the germ for a scene: illusion and trouble; strawberries and truth; strangers, blue, and lights. Now, find your own set of interesting words, and go to work!

QUESTIONS

I emphasized earlier how important questions are to dramatic writing, and this next exercise confronts directly the idea of posing questions at the beginning of a scene.

• • • EXERCISE 8.5

Write a first line of dialogue that asks a question. Write a second line that asks another question. Write a third line that at least partially answers one of the questions but suggests or implies another question or some additional information.

Tom Dulack didn't follow exactly that formula in his comedy *Breaking Legs*, but look how he overlaps questions in the first four lines of the play:

```
Lou:    So what happened?

Angie (opening mail): Nothing happened. What is this,
        469 dollars from Semprini Plumbing?

Lou:    They fixed the toilets in the women's john. So
        what did he do?

Angie: Nothing! Forget it! I thought Semprini owed you
        a favor.
```

Notice how Angie's last line answers one question and implies something else: what's going on between Semprini and Lou that Semprini owes Lou a favor?

Just now I'm looking at the first few lines of some student-written plays, and here are some of the questions that immediately surface:

"How long do I have to stay like this?"

"Oh, my God, did you see *that*?"

"Are you sure it's over?"

"Are you listening to me? Why'd she do this to me? Why didn't you tell me what was going on?"

Notice how such questions not only prompt a response from another character but also engage the interest and curiosity of the audience. Notice, too, how one good question can actually suggest several. Even a simple query such as "Are you sure it's over?" prompts not only "What's over?" but also questions the degree of certainty. "How long do I have to stay like this?" gives us not only the direct "How long?" but also "What is the position?" (which we might be able to see) and "Why is the person in that position?"

So . . . do you think you can handle this now? Then go to it.

LIVE THEATER

Other exercises have asked you to work with dialogue, characters, and plot within the action of the play itself. Through exercises and examples, I have, for the most part, encouraged a generally realistic theatrical style. This exercise, however, encourages you to explore the possibilities inherent in a different theatrical reality: that of live performers in the presence of a live audience.

• • • EXERCISE 8.6

Write a scene that employs in some way the presence of a live audience.

In its simplest form this exercise involves merely a recognition of the audience. Shakespeare's characters regularly address the audience. Tennessee Williams structured *The Glass Menagerie* around a series of Tom Wingfield's monologues to the audience, and Brian Friel also effectively utilizes narrators in several of his plays, including *Dancing at Lughnasa*.

In *Our Town*, Thornton Wilder not only has the Stage Manager speak to the audience, but he plants characters in the audience to ask questions. Going even further, the farcical *Shear Madness* and the musical *The Mystery of Edwin Drood* both ask the audience to determine which direction the action should take, and the scripts come equipped with multiple variations. Such entertainments provide a kind of "interactive" theater in which the audience plays a definite and active role. One of the most elaborate productions of this genre is *Tony n' Tina's Wedding*, in which the audience attends the staged wedding and then proceeds to the reception, where they become full participants in the tacky and wacky festivities.

This exercise will not be everyone's cup of tea, but it does serve as a reminder of the pivotal role of the audience in performance.

ANIMALS

Ramon Delgado, himself a playwright, a teacher of playwriting, and the editor of several collections of short plays, suggested the following exercise, which provides an unusual spur to the dramatic imagination. It starts out by focusing on animals, but the real key is the clever shift to a human situation.

• • • EXERCISE 8.7
Write a scene based on a controversy between two animals.

You should begin by choosing one animal and then asking yourself questions about that animal. What kind of environment does this animal inhabit? What is the age of this animal? What is its physical condition? Its mental state? What are its vulnerabilities? Then try to add some distinguishing personality traits. Perhaps this animal is charming or shy or overbearing. If you have animals as pets or are familiar with animals, you will almost certainly come up with intriguing personality quirks based on the animals you know.

Then consider what animal might be an enemy to the first animal. It could be an animal of a different species or the same species, but you need to imagine its circumstances just as you did for the first animal. Next, look for a situation in which the two animals could be engaged in a controversy. Is the confrontation over territory, food, a mate, or something else? Consider the tactics that each animal uses, and imagine the outcome. Then write a scene developing the dramatic confrontation.

Finally—and this is what really makes this exercise so intriguing—after you've written the animal scene, revise the scene by taking a leap of imagination and transferring the animal circumstances into a parallel human situation. Using this exercise, you may be amazed by what you discover about human motivations.

All of these exercises are designed to get you thinking and working as a dramatist. With your imagination now prompted in a variety of ways, you should be well prepared for the task of putting together a complete play.

9

. . . .

Writing Your Play

EXERCISE 9 • WRITING A PLAY
Write a one-act play. It should be about 20–40 minutes in length. It can be an extension of a previous scene or an entirely new creation. Use correct playwriting format.

After you've tried your hand at the scenes in the preceding chapters, you might be excused for asking, "OK, do I get to write a play *now?*" The answer is: "Yes, of course. In fact, you probably already have!" At the beginning of this book I told you that plays unfold like trees. By working through these exercises you've already begun to nurture your trees.

"But," you protest, "I've only written scenes." What's the difference between a *scene* and a *play*? A scene takes place in one setting and in continuous time. Plays may use different scenes with different settings, or they may use the same setting at different times. Or, a play may consist of just one scene!

The basic elements involved in a scene or a play are the same. A scene and a play both express a story through dramatic means. They both confront characters with obstacles and place them in conflict. They both use dialogue. And, most importantly, they both *animate* the characters. That is, they put them in action. Isn't that what you've done in your scenes?

The difference isn't length, either. Some very complete plays are very short, and some very long dramatic works are very incomplete. The determining factor lies in that word "complete." More specifically, as I indicated earlier, a scene shows us part of a story. A play shows us the *complete story*; it resolves whatever conflicts have been introduced, and it answers whatever questions have been raised.

If you look at the scenes I've used as illustrations in the preceding chapters, you'll see what I mean. In Chapter 7, *Abracapocus* could be summarized this way: A boy tries to avoid writing a paper. That doesn't address the comic atmosphere of the piece, but it does describe the basic action. There is no more to the story. The deed is done, the characters are gone, and the story is finished.

Similarly, *Cliché* in Chapter 6 could be expressed like this: A young man comes to terms with the death of his father through a chance encounter with a homeless man. The piece is, of course, much more than that, but a concise statement helps to focus the significant action of the work.

Cliché, like *Abracapocus*, is a complete short play. It doesn't matter that Chris will surely continue to question his emotions. It doesn't matter whether Chris and Catch ever meet again. A connection was made, and a crucial step was taken. The story is complete.

In *Conversation* in Chapter 3, the problem of the couple's lack of communication lies glaringly unresolved, perhaps even aggravated, but it's unclear if there is more to the story. Is it important and will we ever find out what Nathan is so preoccupied with or why the communication has evaporated? Will Ellen and Nathan ultimately salvage this relationship?

The Donut Shop in Chapter 6 provokes similar questions. Will Emmett ever get his kiss? Will Ceecee get her comeuppance for her flirtatious behavior? There's more to come in that story.

In Chapter 5, the scene between Beth and her Mother represents a complete action: Beth gets her mother's attention and learns a lesson in the process. Unless there is some additional complication, that story is over. So is the relationship between Stephanie and Mike in *Nibbles* in the same chapter. Unless Kurt's appearance adds some new development there would be no need for any additional scene.

In *Big Brother* in Chapter 3 we see the following action: Rick tricks his younger sister into helping him. But there seems to be more going on between the two siblings, and it looks like there could be more to the story.

Good Idea, If It Works from Chapter 4 presents interesting questions. As it stands the scene shows us a young tough pressuring his friend into robbing a store. Is that the whole story? Only the playwright can answer that with certainty. But suppose the whole story includes the young tough double-crossing his friend, and the friend, in deep trouble, avenging himself by killing the cause of his problems. Or suppose the story is about the gullible young man and his father. Imagine the father, disappointed in his son, helps him anyway. Through that act, the young man learns of his father's love and devotion. The play could be a lesson in the consequences of submitting to peer pressure. It could just as easily be a story of parental support or many other stories. The piece certainly leads us to expect that *something* will happen when the theft is discovered.

In other words, to decide if a work is complete or needs additional material, you must first decide *in very concrete terms* just what story you are trying to show, especially where the story begins and where it ends.

Let's assume for a second that you've written an intriguing scene, and you've zeroed in on the story you want to dramatize. Your next question might be, "What parts of the story should I show?" As we saw in Chapter 2, a story can change radically depending on which parts of it you choose to illustrate. The Greeks understood that. The three great tragedians Aeschylus, Sophocles, and Euripides all wrote plays about Electra and Orestes, who revenged themselves by killing their mother, Clytemnestra, and her lover, Aegisthus, because Clytemnestra had murdered their father, Agamemnon, a hero of the Trojan War. Yet the very same characters emerged quite differently in each version of the story because of what the authors chose to show, and those decisions resulted from their understanding of the story.

All of those plays about Electra, however, ended with the deaths of Clytemnestra and Aegisthus—and there is a lesson in that. It's the road the three playwrights traveled to their final destination that makes the plays so intriguingly different. Although at first it seems rather peculiar, many authors need to have an ending place in view as well as a starting place. David Henry Hwang stated just that when he told an interviewer, "I need to know what the beginning of the play is and the end of the play is, and the fun of writing then becomes a way to navigate between the two points." A. R. Gurney said essentially the same thing, commenting, "In order for the story to tell itself . . . I really have to have some idea of where I want it to end up."

The Greeks knew something else, too. They knew that where you begin to show a story is as important as where it ends. They also understood that the starting place, or the "point of attack," as it is sometimes called, derives directly from the point of view of the writer—from what the writer wants to emphasize.

Take the story of King Oedipus, for example. Oedipus is born to King Laius of Thebes and his wife, Jocasta. An oracle tells Laius that this child will murder his father and marry his mother, so the king disposes of the child on a mountainside. A shepherd finds the child and takes him to Corinth, where he is raised in the household of King Polybus. Upon reaching manhood, Oedipus visits the oracle of Apollo at Delphi to divine his future. The oracle warns Oedipus not to return to his home because he will kill his father and marry his mother. Thinking Polybus is his father and Corinth his home, Oedipus runs away to Thebes.

As he approaches Thebes, a man pushes Oedipus off the road, and Oedipus retaliates with a blow that kills the man. Unknown to Oedipus, the man is King Laius, his father. Oedipus continues to Thebes, where he finds the city under a curse. Oedipus saves the city by solving a peculiar riddle, and a grateful city crowns him king. He marries Jocasta. Years later a plague comes upon the city.

All of that is part of the story of Oedipus. But Sophocles does not even *begin* his play until this point, with Thebes under a terrible plague. Oedipus is told that the devastation will continue until the murderer of Laius is found and punished. Oedipus determines to do that, and in the course of the investigation he discovers the horrible truth. In despair, Jocasta commits suicide, and Oedipus blinds himself.

Sophocles could have started the play earlier—when Oedipus is born or when Laius decides to get rid of the child or when Oedipus visits the oracle at Delphi. Sophocles could have pursued the story further than he did, after Oedipus leaves Thebes. In fact, Sophocles did just that, but in a separate play, *Oedipus at Colonus*, which shows the story of the end of the old king's life.

Sophocles could have done any of those things, but he realized that where you start the story significantly determines the story you tell. Because Sophocles focused on the end of this story, the main action of the play might be stated as: "Oedipus discovers the truth."

How do you know where to begin your story? Look for the spots where something changes. In *Hamlet*, the old king is dead. As bad as that is, a kind of status quo has been established. The old king's brother has ascended the throne, and he has married his brother's widow. Shakespeare starts his play on the day that balance is disrupted, when the ghost of Hamlet's father appears to Hamlet and makes him swear to avenge the murder. Just as Sophocles began *Oedipus* on the day the king decides to find a murderer, so Shakespeare began his play on the day Hamlet decides to avenge a murder.

The event that causes a change in the basic situation is referred to in traditional play structure as the "inciting incident." I have refrained from describing typical play structure because I wanted you to explore your own avenues. I didn't want you confined by a rigid system. But now I'd like for you to be able to compare your work against traditional play structure. That might give you additional ideas about how to develop your scenes.

A traditional play is said to have a beginning, a middle, and an end. The beginning starts at a point of balance. The characters and the situation are introduced. Even though other events may have occurred earlier, the contending parties are at a pause. In *Macbeth*, a great victory has just been gained. In *The Glass Menagerie*, an uneasy truce exists between the family members.

Then a new element is introduced. This is the *inciting incident*, and it propels characters to action. Often the inciting incident involves the arrival of a new character. In *Dancing at Lughnasa*, the family has just purchased a radio, and Uncle Jack has just returned from 25 years' work with a leper colony in Africa. Most importantly, Gerry Evans, the narrator's father, returns. In *Hamlet*, the ghost arrives, demanding revenge.

The inciting incident can occur in other ways as well, but it always includes some new information. In *Macbeth*, the witches incite Macbeth's thought of the crown, and then the king arrives at Dunsinane Castle. Teiresias tells Oedipus that he must find Laius's murderer. In *The Glass Menagerie*, Amanda discovers that Laura has deceived her about attending typing school, which focuses Amanda's attention on finding a husband for Laura.

The inciting incident sets in motion the desires of the main character to accomplish a goal—Macbeth to become king; Hamlet to avenge his father's murder; Amanda to procure a settled life for her daughter. Author Lillian Hellman told aspiring writers that a playwright has about eight minutes to let the audience know who the play is about, what is at stake, where the play is going, and why. That's the beginning.

The desire of the main character meets some form of resistance, and, in the middle section, the play proceeds through a series of obstacles and conflicts. This is sometimes referred to as the *rising action*. Macbeth overcomes his own fears and hesitations. Hamlet tests his uncle's reactions to a play. Amanda pressures Tom to bring home a boy for Laura.

Within the middle part, good playwrights find ways to raise the stakes, to increase the emotional and psychological pressures on the main characters, and to complicate their lives. Not only does Macbeth himself want to be king, but he also receives passionate pressure from his wife. As Hamlet pursues revenge, he feels the additional pressure of a deteriorating relationship with his beloved Ophelia. The gentleman caller whom Tom brings to the house turns out to be the one boy Laura has always loved—and he's engaged to someone else.

Often, as we saw in *Big Brother* in Chapter 2, action heads in one direction and then suddenly shifts. One side appears to be winning, only to suffer a setback, which is called a *reversal.* Just as it appears the boy will get his sandwich and the girl her ride to school, Rick double-crosses his sister. Hamlet stalks his uncle to his chamber, but changes his mind when he finds him praying. Laura kisses Jim O'Connor but then discovers he's going with another woman.

Finally at the end, one side wins or loses utterly and decisively, and the battle is over. Hamlet kills the king, and he and several others die; Fortinbras is left to pick up the pieces. Macduff slays Macbeth, and Malcolm is crowned the rightful king. Amanda's far-fetched dreams for Laura are shattered like the glass figurine, and Tom leaves town.

WHY NOW?

As you develop your play, always ask yourself: "Why do these characters do what they do *now?*" Most of us procrastinate. We would just as soon put off action, particularly important, decisive, and difficult action. Characters are the same way.

I will often ask a writer, "Why does this character erupt—or do whatever he or she is doing—just now?" Frequently the answer comes back, "Things built up." I understand that "things build up," but the character didn't erupt the day before. Or the day after. What caused the disturbance on *this* day?

Think about the expression "the straw that broke the camel's back," which means that a camel—or a person—can carry a very heavy load, but eventually a limit will be reached. Physically or psychologically one more item, one more tiny piece of straw will break down the animal.

We, the audience, can't see all the problems that build up, but we definitely want to see that last piece of straw, and you, the author, have to identify it for yourself and for us. In many cases that moment is the beginning of your play.

Whenever possible in the beginning and middle sections of your play, put your characters under pressure to act. If John wants to ask Barbara to marry him, we have a typical situation. If John knows Barbara is also dating someone else, there's more pressure for him to speak up. But if John knows Barbara's leaving tomorrow for a job in another part of the country, then John, if he's going to pop the question at all, has to do it *now.*

Why does Macbeth kill his sovereign, Duncan? The prospect of becoming king tantalizes him, to be sure. His wife urges him to it. But beyond the desire and the encouragement, he is presented with *opportunity* and a *time limit*. The king comes to stay at Dunsinane. If Macbeth's going to do the deed, he has to do it <u>now</u>. Always ask yourself, Why does the play take place on *this* day rather than another day? In *The House of Blue Leaves*, for instance, John Guare sets the play on the day the Pope visits New York.

I said previously that as you develop your play, as you identify the story, figure out where to start it, and create the pressures for action, the themes and ideas of your play would emerge. Plot details and thematic concerns reinforce each other; the point of the play is inextricably bound up with the selections you make regarding the details.

As we saw earlier, *Good Idea, If It Works* could move in any of several very different directions. It depends on the story the author wants to focus on and what the author thinks is important. In other words, the story comes from the author's sense of values. Authors must be willing to examine the most compelling concerns, for without a sense of values a plot is merely a mechanical contrivance and characters merely mechanisms for action.

Often the themes of a play emerge in details that appear almost incidental to the action of the play. Lyle Kessler's *Orphans* features a shadowy character who seems to have gangland connections. He enters the house of two underprivileged orphans and changes their lives. He helps them procure jobs, fine food, and natty clothes. More importantly, he provides emotional support by giving them "an encouraging hug," which becomes a central visual motif of the play. When the man dies, the orphans try to place the dead man's arms around their shoulders.

The visitor sings snatches from a particular old song, "If I had the wings of an angel." Although the man sings only a few lines, some of the lyrics of that song refer to poor sinners being enfolded by those angelic wings. In dialogue, too, the theme is expressed. After one of the young orphans is given a map of his neighborhood in Philadelphia—not accidentally the "City of Brotherly Love"— the boy declares, "I know where I am now." Indeed, this "lost boy" has now found himself.

The details of this play might have been different. The stranger could give the boys a pat on the shoulder. He could sing another song. The orphan could speak other words. The setting could have been another city But Kessler chose to dramatize the theme of encouragement through the details he selected.

STYLE

Another element that emerges as the play develops is style, which means the manner in which the play is done. Few authors set out to write a play in a particular style. Rather, the style of the play evolves from the material itself. In *Orphans*, Kessler wanted the realistic trappings of a seedy Philadelphia apartment. In *Our Town*, Thornton Wilder wanted to show us that we should, for our few hours on the stage of life, try to see and appreciate everything around us. To achieve his point, Wilder stripped the stage of scenery and props so the audience would be forced to imagine all the shapes, colors, textures, sounds, and tastes of life with which Wilder stuffed the play. The style of *The Donut Shop* in Chapter 6 suggests a realistic setting, while *A Man in a Bus Terminal* in Chapter 1, despite the realistic setting, evokes a rather abstract world through the exaggerated characters.

In line with that, I offer a warning regarding overt abstraction and abstract characters. It is easy to see good and evil in the world, and inexperienced writers are tempted to put those qualities on the stage as characters. Trying to show a young man in conflict, the author will give us "Angel" and "Devil" presenting their cases in the boy's ears. Or "Good" and "Evil" will put in appearances. Seldom does that kind of abstraction work very well. Just as the essence of the character is defined in a name, such characters usually come across as predictable, one-dimensional voices. They lack complexity and the human qualities that most attract our dramatic attention.

Abstractions are generally more successful when conceived in human terms. In the *Oh, God* series, for instance, George Burns proved endearing because of his idiosyncratic human foibles. The stranger in *Orphans* is a sort of angelic fairy godfather who transforms the boys' clothing and environment just as surely as Cinderella's fairy godmother transformed objects with her wand. But we care about the visitor because of his human interactions with the boys.

Now you're fully equipped to put together your own play, so lets proceed to Exercise 9.

• • • EXERCISE 9.1 Writing a Play

Write a one-act play. It should be about 20–40 minutes in length. It can be an extension of a previous scene or an entirely new creation. Use correct playwriting format.

Before you begin your own piece, you might want to read this play by one of my students.

CABLE MAN
By Jerome Hairston

THE SETTING is a rundown one room apartment in downtown Newport News, Virginia. There's a table and some chairs. A bed. A phone. Scattered bills. An old TV with a cable box. A baby crib has been constructed out of two chairs facing each other with a large plastic storage bin resting between the chairs. Some blankets and clothes can

be seen in the bin In the tiny kitchen
area the oven door is open, and a fan sitting on the oven door tries valiantly to
push some heat into the room.

The television is on. Cable infomercial. A
young woman, LADY, watches. She's 19,
white, dressed in an oversized "Garfield"
nightshirt, faded blue jeans, and an old
Army jacket. There's a knock at the door.

LADY

Hello?

VOICE (outside door)

Yeah.

LADY

Hello?

VOICE

Sorry, uh, 302. I'm looking for, is it . . . Yeah, 302.

LADY

Hello? Who? Who?

VOICE

Nation. I'm from Nation Cable.

LADY

What? I don't . . . What are you doing?

VOICE

I'm knocking. It's the cable man, lady.

LADY

I didn't . . . I didn't call.

VOICE

I know you didn't. I'm about to disconnect
          (Knocks.)
I need to talk to you.

LADY

          (Rushes to the door and opens it. HARRIS, a
          31-year-old black man is visible at the
          door. He wears a work shirt with a name
          tag, work boots, and tools in a work belt.
          He carries a clipboard.)
Disconnect my . . . I didn't call.

HARRIS

I know this. I'm not here on a call.

LADY

I didn't.

HARRIS

There's a matter of payment.

LADY

I do pay. I didn't call you.

HARRIS

I know you didn't. In fact, we called you, I believe.
To talk about your bill.

LADY

I said I . . .

HARRIS

Yes, I'm sure you've paid, but . . .
          (Consults clipboard)
says here not since October. They were supposed to call
you to tell you that. . . .

LADY

That's not my fault. The phone don't work. It don't. It
don't ring. No tone. It don't work.
          (Goes to phone to show him. HARRIS comes
          into the apartment.)
Check it. Listen. Go ahead.

HARRIS

Look, they were supposed to call, and if you ain't get
the call, you still should've got a notice in the mail.

LADY

The mail? The mail? Things get lost in the . . . That's
not my fault either. You want to blame me for that?
For that? Don't start blaming me for things I don't got
control over. Point your finger somewhere's else. Disconnect someone else's cable. Don't come . . .

HARRIS

But I'm here, Lady. And I'm here to disconnect. You
don't want to be disconnected. Cool. Let's sit. We can
talk. But I can't do a thing. . . .

LADY

Talk?

HARRIS

Yeah. I'm sayin' me and you can solve this.

LADY

Don't disconnect my cable. You can't . . .

                                HARRIS
Okay. Look, I can help you.

                                 LADY
Help me?

                                HARRIS
Yeah. What I can do . . . Can we sit?

                                 LADY
You want to sit?

                                HARRIS
Yeah. We both can sit. In chairs. Me and you. I just
figured it would be more comfortable.

                                 LADY
You're not comfortable? I'm . . . I . . . What am I doing?
I'm making you uncomfortable is what you're saying?

                                HARRIS
I just think it would be better if we sit. Face to
face. And straighten this out.
                    (Pause.)
So?

                                 LADY
So.

                                HARRIS
Can we sit?

                                 LADY
You . . . I . . . Yes, but you have . . . you got to be quiet.
There's a baby here.

                                HARRIS
Understood.

                                 LADY
And she's sleeping. And if you wake her, she'll . . .
she'll . . .

                                HARRIS
I hear ya.

                                 LADY
I mean she cries. Bawls something terrible. Once she
starts . . .

                                HARRIS
I got it. No disturbances.

                    (Cautiously, they sit. LADY immediately pops
                    up.)

LADY

Don't say anything. I know.

HARRIS

Know, Lady?

LADY

The smell. You was about to say it stinks in here.

HARRIS

No, I . . .

LADY

It might smell shitty in here, but you got to remember there's a baby in this place. Baby's . . . shit. I mean, maybe I shouldn't say it like that, but, you know, they do. And they spit up. They . . . they smell bad sometimes. So don't blame me. I keep this place clean. It might stink in here, but it's not dirty.

HARRIS

I understand. It ain't dirty.
        (Flips through pages on clipboard.)

LADY

I don't got much, but I do what I can with the little I got. You like my baby's crib?

HARRIS

        (Gives a quick glance.)
Yeah, nice.

LADY

You got to look.

HARRIS

I see it.

LADY

She sleeps good in there. She dreams. Look at her.

HARRIS

        (A bit longer glance at the crib.)
Mmm. Cute.

LADY

Made it from nothin'. May not look like much, but she likes it. Figure that's what's important. Bet you most other baby's would like it, too. People pay all that money for that store bought stuff. I should sell cribs on my own. Probably make a fortune, huh? I mean, I could be good at doin' somethin' like that.

                    HARRIS
Let's get started.

                     LADY
Yeah, I could be good.

                    HARRIS
            (Takes a sheet from clipboard.)
Here. Take a look at this.

                     LADY
What's that?

                    HARRIS
Over here. This is a form. See?
            (LADY sits warily.)
You sayin' you're havin' problems with payment, right?

                     LADY
I never said that.

                    HARRIS
Well, in case you do, this here is a form that'll . . .

                     LADY
Don't disconnect my cable.

                    HARRIS
That's what I'm talking about here. This'll take care
of that.

                     LADY
Can't you take care of that? I mean, you're the one
with the tools. The ones on your belt. You're the one
with the . . . the power to do . . . you know . . . whatever.

                    HARRIS
But see, these forms . . .

                     LADY
You don't need forms, do you? What's wrong with asking?
I'm asking you now. You have the power, so I'm asking
you. Don't cut it off. You can do that. Cause I . . .

                    HARRIS
Lady, hold on. . . .

                     LADY
. . . cause I asked. That how it should work. Please
don't disconnect my cable. I said it. I'll say it again
if you don't hear me . . .

                    HARRIS
I hear you fine, but . . .

                              LADY

. . . cause sometimes people don't hear. When I tell
them, look, you can't, please don't, they still do.
They say I don't pay, but I do. I send them all I got,
but they, they call on the phone, and I tell them and
they don't listen. They just . . . there's a bad line
somewhere or something, some problem with the connection, cause they don't hear me, and I ask. . . .

                              HARRIS

Listen, Lady, I'm not here to . . .

                              LADY

I do. I'm polite. I say "please don't cut me off,
don't cut me off," but things stop. Like the phone.
It just stopped. Stopped working. So now no one calls.
Nobody listens. I mean, are you? Are you listening to
me?

                              HARRIS

Yes, I am.

                              LADY

Then don't. Don't take away my cable. If you heard me
asking, then don't.

                              HARRIS
                  (Raising his voice to talk over her.)
What I've been trying . . .

                              LADY
                  (Motioning to crib.)
Shhh!

                              HARRIS
                  (Lowering his voice.)
. . . trying to explain to you is that I'm not here to
cut you off.

                              LADY

But you said . . .

                              HARRIS

Wait. Just wait. I'm supposed to be here to disconnect,
but we can work something out. I understand what you're
going through.

                              LADY

You do?

                              HARRIS

Yes. I know what it's like to hit hard times. If you

ain't hit hard times once in your life you ain't human.
You a human being, right?

> LADY

Yeah. I guess. Yeah.

> HARRIS

That's right. You can run into snags just like everybody else. What I'm saying is that we can help you.

> LADY

We?

> HARRIS

Nation. Nation Cable. And we'll try to work things out
with you. You ain't the only one who gets caught like
this. In a . . . situation? So what Nation Cable has done
is come up with a system.

> LADY

System?

> HARRIS

A brand new system to help people like yourself. A
policy. To show people that we're not out to be the
bad guy. We've done this with plenty of other folks in
the . . .

> LADY

Other?

> HARRIS

Yeah. Other folks in the neighborhood.

> LADY

You say this to them.

> HARRIS

What?

> LADY

What you're saying now. To them.

> HARRIS

Yes.

> LADY

And you're saying it to me.

> HARRIS

That's right.

> LADY

You say this a lot?

                                HARRIS
I'm just . . .

                                LADY
Every day. To other people. You say this. Say it so
much you got it memorized or something. Like this is
all planned. Like a con. Well I don't want to be
conned.

                                HARRIS
                    (Raising voice, then looks to crib and
                    catches himself.)
I'm just letting you know . . . letting you know this is
nothing to be ashamed about. That it's not about you.
We do this with a lot of people.

                                LADY
I'm not . . .

                                HARRIS
We're saying it's all right.

                                LADY
I'm not like other people in this . . . They're different
from . . . Look. Look, I got cheekbones.

                                HARRIS
Excuse me?

                                LADY
I got cheekbones. Nice high cheekbones, you know, like
on . . . My face don't belong here.

                                HARRIS
Okay. Well, what I got here is a form.

                                LADY
What? You don't believe me?

                                HARRIS
I'm sure they're fine.

                                LADY
I can prove it to you, if you don't believe me. I can
show you.

                                HARRIS
That's okay.

                                LADY
No. No, you want to see? Feel. . . .
                    (LADY grabs HARRIS'S hand and places it to
                    her cheekbone. HARRIS is stunned by the

gesture. LADY looks down as though trying
to feel her own cheekbones through HARRIS'S
touch.)

See what I mean? I'm . . . It's different. See?
(LADY looks up. She sees the shock in
HARRIS'S eye and lets go of his hand.
Silence.)

HARRIS

What I was saying is, that this is a form.

LADY

I?

HARRIS

What?

LADY

You said, "I was saying."

HARRIS

Yeah.

LADY

I thought it was "we."

HARRIS

Well, I mean me, then. I'm trying to show you that
there's something good that's going on here. You want
to keep your cable, am I right?

LADY

Yes.

HARRIS

All right. We can do this. What you got to do, is fill
out this form, and I'll take it from there.

LADY

Form?

HARRIS

This one. Right here.
(Gives form and pen to LADY, who receives
it as though it might bite.)
You just write down your name and everything here, and
then on the space there, you explain why . . .

LADY

Why you shouldn't cut off my cable?

HARRIS

Something like that.

LADY

I'll tell you why you can't take it away.

HARRIS

Write it down.

LADY

But I'm telling you.

HARRIS

It don't work like that. It's no good unless you write it down.

LADY

But I'm telling you. What's wrong with just . . .

HARRIS

They, down at Nation, need to know to . . .

LADY

I thought I was talking to you?

HARRIS

You are, but they have to process this before anything can happen.

LADY

But I don't understand why you can't just . . . Go down there and tell them. Tell them I pay. Just leave my cable the way it is, and tell them. Why can't you just do that?

HARRIS

Cause I'll lose my job, Lady. You know what I'm sayin'? I'm tryin' to help you here. You just need to let me. Now c'mon, just write what you . . .

LADY

I don't want to.

HARRIS

Look, it doesn't take much to just pick up the pen and . . .

LADY
(Vicious.)

I DON'T!
(Looks to crib and lowers voice.)
I . . . I'd rather just tell you. That way I know some-body heard. Somebody was listening. Just let me tell you. Please. I just . . . I don't want to write it, that's all.

                              HARRIS
You can't do it, can you?

                              LADY
I can't? What? You think I can't. I graduated, mister.
I did. Got a diploma sitting right over there. Signed
by the principal of Danby High School. You want to see
it? I'll show it to you, if you want to see it.

                              HARRIS
You want me to write it for you:? I can do that, you
know.

                              LADY
You're not listening. I told you, I graduated.

                              HARRIS
I know. I know you did. But do you want me to write it
for you anyway?

                              LADY
I don't want to be judged, you know. There's a lot of
things I can do. I mean, look over there. That crib. I
did that. Me. By myself.

                              HARRIS
I know you did. I ain't judgin' you. I just want to
get this settled. I'm sure you want to get this
settled, too, right? So you talk. I'll write. Sound
good?

                              LADY
You really got to listen. Listen hard. And write down
everything I say.

                              HARRIS
Fine.

                              LADY
Everything. No making shit up. You have to listen to
every word I say.

                              HARRIS
Let's get going.

                              LADY
You serious about doing this?

                              HARRIS
Yeah, I am.

                              LADY
Why?

> HARRIS

I've done it before. Ain't no big thing.

> LADY

I'm . . . I'm sorry I'm givin' you a hard time. I shouldn't be acting this way to you . . . mean. You . . . you're a good person, aren't you?

> HARRIS

I'm just the cable man, Lady.

> LADY

You really want to listen?

> HARRIS

Yeah. Talk to me.

> LADY

It's the only way she sleeps. My little girl. The only time she closes her eyes. Where she's quiet and still. Trust me, you don't want to be around when she's not. She starts up cryin'. She gets hungry sometimes, you know. And I try to tell her things. I want to tell her that I'll be gettin' a job soon, so she won't have to worry about nothin', but I don't think she'd understand that. She's just a baby, you know? So I just tell her, "Mommy'll be gettin' you some crushed peaches soon. Don't that sound good? Some crushed peaches just for you, baby." She'd like that, don't you think? You writin'?

> HARRIS

Yeah, I'm gettin' it down.

> LADY

I keep telling her and telling her, but the tears just won't stop. But I turn on the TV. And, finally, she calms down. And every time she starts up again, I just flip the channel. You know, cable, there's so many. You can flip all day if you wanted to. I don't know, must be somethin' about the voices, helps her sleep. And as she's sleeping I tell her about the people I see. People on the screen. About their cheekbones. About how Mommy got cheekbones just like that. That she can be on TV if she wanted to. That she could be anything if she wanted to. She likes hearing about that. I know cause sometimes I look over at her . . . dreamin'. Dreamin' about Mommy.
> 　　　(Pause.)
> But everybody tries to wake her. The heat gets shut

off, the rats keep crooping around her crib—the crib I
made for her—and now, you, you come and try to take
away our. . . . My baby needs it, you see. I stand on
line, on line with them people, knowin' damn well I
don't belong there. I don't. It's a bunch of faces that
ain't anything like mine. They stare. Eyes popping from
their skin. But still, I stand there, humiliated, and
I, I get my checks and I pay. Believe me. I pay. So I
don't understand why they sent you. You . . . you got to
help me.

                    HARRIS
Okay, but look, I want to help you.

                    LADY
So you won't cut off my . . .

                    HARRIS
Wait, now. Hold up. I want to help you. That's why I'm
gonna tell you this. I mean, I see kids like you just
about every day.

                    LADY
I told you I'm not like . . .

                    HARRIS
Every day, I see kids like you. And I can't help feeling, you know, you shouldn't have to live like this.
You got a kid over there. A little baby girl. And you
say she's hungry, right?

                    LADY
What're you tryin' to say? You saying I don't care for
my kid. That what you're saying? I keep her from crying, Mister. I do. Unlike most mothers in this building. The sound will keep you up all night, the cryin'
babies in this place. But my girl dreams. She dreams,
okay?

                    HARRIS
I know you're trying. But there's things you can do.
For yourself. For your girl. Look, you say you get
checks, right?

                    LADY
That's right. And I pay. I'll pay you.

                    HARRIS
          (Erupts, then regains control.)
I don't want you to pay me! Look, it's only cable,
Lady. Can I give you some advice?

                              LADY
No. You can't.

                             HARRIS
Now just hear me out. I mean you should do something
good with your money. Use those checks to buy your baby
some food. Some of them crushed peaches, huh? Cause to
be honest with you, this setup you got goin' here,
ain't gonna last long. The electric is gonna come after
you next. It's December out there, you know. You really
need to get the heat turned back on in this place.

                              LADY
My baby's warm. She gots blankets, she got almost all
the clothing I own wrapped around her body. I do it
cause she's cold. If someone's cold you keep them warm.
If someone is cryin' you do what you can to make it
stop. You don't think about yourself. I do everything I
can to . . . You gonna keep her warm?

                             HARRIS
I'm just tryin' to help you. I'm trying to help you
make things right.

                              LADY
You want to help me make things right?

                             HARRIS
Yes.

                              LADY
Then get out of here and leave me, my baby, and my
cable alone.

                             HARRIS
Look . . .

                              LADY
Get out!

                             HARRIS
Calm down. Come on, you'll wake the kid.

                              LADY
I said get out of here.

                             HARRIS
Listen. I'm . . . I'm sorry, okay. It was none of my
business. You're doin' just fine here, I see that. I
shouldn't have said nothin'. I'm sorry.

                              LADY
I'm trying my best.

                              HARRIS
I know you are, Lady. Listen up, I'll do what I can.
For you. Look, I wrote down what you said and I'll
take it down to Nation and you'll probably get an extension, okay?

                              LADY
You sayin' I get to keep my cable?

                              HARRIS
Well, yes. Yes, you will. It'll take some time to process, but, yeah, you'll keep your cable.

                              LADY
I get to keep my . . .

                              HARRIS
. . . your cable, yes. I'll work on it.

                              LADY
You'll do that.

                              HARRIS
That I will.

                              LADY
You're gonna make things right. Gonna make it right.

                              HARRIS
I'm gonna try.

                              LADY
              (After a pause.)
I'm sorry. I'm sorry I . . . Thank you.

                              HARRIS
Not a problem.
              (He moves toward the TV.)
Now what I'm gonna do now is unhook your box here,
then I'll be out of your way. Now it's nothin' . . .

                              LADY
What?

                              HARRIS
Nothin' to worry about. It's just somethin' I gotta do.

                              LADY
Why?

                              HARRIS
              (Skews the TV so it faces the audience and
              begins disconnecting.)

It's just procedure. You'll get it back when you're
reconnected.

LADY

Reconnected? But you said you . . . You saying you lied
to me?

HARRIS

I didn't lie.

LADY

Reconnected?

HARRIS

It won't be off for good. A week or so, then I'll be
back to reinstall.

LADY

Weren't you listening?

HARRIS

Yeah, I heard every word, and I'll tell them downtown.
Don't worry.

LADY

Don't worry? But my girl. She'll start. I can't be
waiting for a week.

HARRIS

It might be even sooner.

LADY

You can't. You can't do this.

HARRIS

I have to, lady. Look, it's no big thing. It'll be
back.

LADY

You don't understand what you're doing. She'll . . .

HARRIS

You don't have to tell me about babies, Lady. I got
two of my own. Yeah, she might cry, but believe me,
they can't keep it up for too long.

LADY

Please.

HARRIS

Trust me. Nobody can cry forever.
          (The cable is disconnected. The television
          fuzzes. LADY and HARRIS look to the crib.

The room is silent. HARRIS begin gathering
his things.)
See? What I tell you.

LADY

She . . .

HARRIS

Look at that. She's a better kid than you thought.

LADY

Yeah, I guess she's full of surprises. She can be a
good girl when she wants to be. She can be the sweetest, prettiest thing in the world. You know, people
don't stop and say a damn thing to me usually, but
when I got her with me, totally different story. They
look at her, and, you know how it is, "so cute, so
adorable." Even say she looks like me. The eyes, the
face, you know, the cheekbones. It's funny. Usually,
nobody ever pays attention to my eyes, my face. But if
I'm holding her, all the sudden, the world sees. Sees
me. It's almost like I ain't here, ain't alive unless
she's with me. She is precious.
            (During this speech HARRIS makes notes on
            his forms, then walks to the crib and looks
            in. He takes a closer look. Then he puts
            down the clipboard and cable box and re-
            turns to the crib.)
Especially when she smiles. She smilin'? That what
she's doin'? She smilin' in her sleep? Mister?

HARRIS

No.
            (Pause.)
We should call somebody.

LADY

What?

HARRIS

I'll call.

LADY

Call? Why do you need to call anybody?

HARRIS

            (Goes to phone.)
Somebody's comin' to help, okay? It's gonna be all
right.

                              LADY
All I asked was if she was smiling. Smiling so you
could see.

                              HARRIS
                    (Realizing phone is dead.)
Dammit.

                              LADY
See why you can't take this away from her.

                              HARRIS
Look, Lady, I have to go to the truck.

                              LADY
                    (She's blocking his way to the door.)
Go? Wait.

                              HARRIS
Please. I need you to be calm. I need you to listen to
me.

                              LADY
Look, I'll do anything. Just leave things like they
are. I'll pay. I'll catch up, I promise.

                              HARRIS
I need to go to the phone in my . . . I . . . Listen, I'm
trying to help you.

                              LADY
What else do you want?

                              HARRIS
I don't want anything.

                              LADY
Then why are you doing this?

                              HARRIS
I know you're upset. But don't do this.

                              LADY
Why do you have to take this away from her? She's just
a baby girl. She's maybe quiet now, but it'll start
again. It always starts again.

                              HARRIS
Trust me, I'm trying my best here to . . .

                              LADY
Trust you? After all your lies? All your damn lies?
                    (Advances on HARRIS.)
You're evil. That's what you are. Evil.

(Shoves HARRIS.)

HARRIS

Now look, get a hold of yourself. Wake up, Lady. You understand what's going on here? Do you?

LADY

Understand? I live here. I live here you son of a bitch!
(More shoves.)
This is my home! You can't come into my home and do this to me!
(Hitting him.)

HARRIS

Stop this!

LADY

You have no right! No right!
(Swinging at him.)

HARRIS

Stop it! I mean it!

LADY

Oh, you mean it? Just like you meant the rest? You still gonna make this right? That's what you told me wasn't it? You gonna make this right?
(Hits him again.)
Huh? So do it. Do what you said.
(Hits him.)

HARRIS

Stop it or I swear . . .

LADY

(Hitting him with every line.)
Make this right. You lying piece of shit! MAKE THIS RIGHT!
(HARRIS viciously pushes her back and she falls to the floor. She looks at him, then, softly.)
Make this right.

HARRIS

(HARRIS looks at the feeble body in front of him and inside his even more feeble soul. Pause.)
It's not my job, Lady.
(HARRIS gathers his gear.)

I'll call. From the truck. Somebody will be here soon.

There's nothing else I can do. I wish there was something I could say, but I'm just . . . Your cable's been
disconnected. That's bout all I'm qualified to tell you.
I'm sorry.

>   (HARRIS heads to the door.)

>   LADY

Hey Mr. Cable Man?
>   (HARRIS turns.)
Are you cold?

>   HARRIS

No. I'm not.

>   LADY

That's not the reason I asked.

>   (Silence. HARRIS EXITS. LADY wraps herself
>   in a blanket from the bed and moves to the
>   television. She kneels before the televi-
>   sion. Framed by the glow, she touches her
>   fingertips to the scrambled gray screen as
>   the lights dim, leaving her supplicating
>   figure tinged by the light from the
>   screen.)

>   THE END

EVALUATION

From one perspective, this play focuses on a young woman utterly
ill equipped to cope with the daily activities of life. But it also focuses on
the relationship that emerges between the woman and Harris, the cable
man. Although he indicates that he is just doing his job, he is clearly
touched by the woman's plight and by her simplicity. He makes a connection, but that connection is shattered when he discovers the problem with
the baby and he retreats to his businesslike facade.

As with many plays, the arrival of a new character instigates the
action. Harris's first obstacle is just to get inside the apartment. Then he
tries to soothe the woman, to get her to sit, to talk, to relax. The play
reaches its first reversal when Harris attempts to explain the company plan
to the woman. That tactic backfires, however, when the woman suspi-
ciously dismisses his spiel as a "con." Harris then prods the woman to
write out her problems, and that leads to his discovery that she is illiterate.
Her insistence that she has a diploma may or may not be true, but it certainly reveals her defiant pride. When Harris agrees to write for her, the
woman for the first time begins to accept him. "You're a good person," she
says, apologizing for her outburst.

The two characters in "Cable Man" not only talk differently, they
think differently. Harris is very organized, and his language proceeds

carefully, logically from one item to another. The woman's mind and her language make huge jumps, as when she leaps from "You say this a lot" to "Like a con." As Harris struggles to reestablish trust, the woman produces one of those startling mind jumps when she says, "Look, I got cheekbones." That thought associates her with the television images she idolizes, and it leads to the riveting moment when she presses Harris's hand to her face to show him her cheekbones. That touch creates too much intimacy for Harris, and he withdraws to his official language: "What I was saying is, that this is a form." The young woman's simplicity continues to arrest Harris and the audience, however, for she notices the subtle difference in Harris's speech: "You said 'I was saying' . . . I thought it was 'we.'"

The author has identified the characters by race, but race plays little direct role in the play. Certainly the woman might have race in mind when she says, "I'm not like other people in this . . . My face don't belong here." A difference in the race of the characters could also augment the gap between them and the difficulty they have in making a connection. Still, it seems that the situation could bear performers of any race.

Eventually, moved by the Lady's monologue about her child, Harris discards his businesslike demeanor and tries to give the woman some heartfelt advice. But just as he recoils from her touch, so she flies from his ministrations. Again Harris reassumes the persona of his job, which leads to a major reversal. Just as the woman is finally convinced that she will keep her cable box, Harris goes to remove it, destroying her shaky trust. After Harris unplugs the box, an uneasy calm prevails until Harris peers into the crib and makes his fateful discovery.

In an earlier draft of the play, Harris actually announced, "Your baby's dead," but then the focus of the play became that single fact rather than the characters and their relationship. By keeping the problem implicit rather than explicit, the author retains the emphasis on the two people and their interaction.

Once you've written your own play, you will want to examine it as objectively as you can, just as I have done with the scenes in this book. When you're ready for a self-evaluation, begin by asking yourself these questions:

— Who is the main character? Who is the play about?

— What does this character want?

— Does this character undergo any change in the course of the play?

— Why does this action happen on this particular day?

— What is the most important moment—the climax—of the play?

— Why should we, the audience, care about what happens to the characters?

— Is the setting significant to the action, and is the action used to reveal the characters?

— Can you summarize the action of your play concisely, as in the following format:

This play is about a [PERSON: for instance, "a woman," "a man," "a boy"] who [ACTION PHRASE OF A FEW WORDS: for instance, "wants to be king," "seeks revenge for his father's murder," "pursues the truth of an unsolved murder"].

If we apply these questions to *Cable Man*, some of the answers are rather complex. While the woman is apparently the main character, she is in fact only reacting to her situation. It is Harris's desire to remove the cable box that forces the play. In other words, in the formulation I've suggested, "this is a play about a man who removes a cable box from the apartment of a destitute young mother." In the course of that action Harris makes profound shifts and connections, from businessman to friend and back again.

The action happens *now* because Lady hasn't paid her bill, and this is the day Harris must disconnect the box. One climactic moment occurs when Harris gets what he wants; he successfully disconnects the box. But a second climactic moment occurs when he sees the baby. His dreadful discovery undermines his successful *dis*connection of the cable box and amplifies his unsuccessful attempts at connection with the woman. It leads to their final violent confrontation and his difficult exit. Although he has the cable box, he appears more defeated than victorious by this confrontation with life.

I believe the author has given us characters we can care for, which, especially with the girl, is a challenging task. Like Harris, we want to shake her at times. Yet we see and understand her pride. She seems genuinely concerned about her baby. Her flights of fancy are exhilarating, and the moments when she lets down her defenses are touching.

Not only does the author include superb details about the room, such as the fan trying to blow heat from the oven, but he integrates set pieces and props. The crib, the television, and the cable box, are, of course, central. The phone, mentioned at the start, reappears as an obstacle at the end. The doorway itself becomes a hurdle at the beginning and again at the conclusion. Props and actions mesh to reveal character, as when Harris gives Lady the form and the pen and when he finally disconnects the box.

If you have trouble answering those questions about your play, then you need to focus your action more carefully as you work on your second draft. And then it's on to your finished script. You've placed the characters in conflict, and you've put them under pressure. You've included significant details. You've explored the human qualities of your characters. And out of that grows your play—complete, powerful, intriguing, sensitive, witty. Your play will astonish the audience. It will entertain them, and it will move them to laughter and to tears.

10

• • • • • • •

Marketing Your Play

So you've written your first play.

Then what?

A play isn't meant just for you. It's meant for other people. So the next question is, What do you do with your play once you've written it? The fact is, marketing your play is at least as time-consuming and demanding as writing your play. It can also be far more frustrating because while you can control the writing, you can't control what other people think of it—or how long they take to respond. In addition, marketing your work can be quite expensive. In other words, you'd better get used to the idea that you're going to work just as hard marketing your play as you did writing it. You do, however, have one huge advantage in finding outlets for your play that no one else has, and that advantage is YOU.

FORMAT

After you're convinced that your play is as good as you can possibly make it, the next thing to do is to make it *look* as good as you can possibly make it. Almost everyone now uses a word processor of some sort. Your material doesn't need to be fancy, but it should be *clean*. That means a good printer. It means use that spell checker, and pay careful attention to the homonyms that most spell checkers don't find, such as "it's/its," "your/you're," and "their/there/they're."

You should follow the playwriting format illustrated in the scenes in this book, with top, bottom, and right-hand margins of at least one inch and a left-hand margin of one and one half inches to allow for binding. Plays should be spiral, velo, or tape bound on the left side. Three holes with metal brads will give your play an old-fashioned appearance and a negative first impression. For a one-act play you would use consecutive page numbering in the upper right-hand corner. If the play has more than one act, pages are numbered by act (e.g., II–12). Acts are given capital Roman numerals, and scenes are written with small Roman numbers (e.g., II–iii–17).

A title page should include the title of the play, a brief designation such as "A Comedy in One Act," and the author's name. These should be just above the center of the page. At the bottom right is the playwright's address, and in the bottom left the copyright notification would appear.

A copyright provides essential protection for your ideas, and, fortunately, it is easy to secure. Authors may copyright unpublished and unproduced plays by writing to the Copyright Office in Washington, D. C., and requesting copyright forms.

A second page would include the cast of characters. Each character should be listed, followed by a brief description. The "setting," listing the time and place for each act and scene, should also be included on this page.

Anytime you send out a play for consideration, you should include a SASE (self-addressed stamped envelope) for the return of the script. You'll also need to include a cover letter explaining your reason for sending the script. It's also a good idea to include a return postcard for acknowledgment of receipt of the script.

Copying, binding, and mailing out multiple copies of a script is expensive. Copying and binding just a single copy of a 20 to 30 page one-act play might cost five dollars, and mailing with return postage and envelopes may run almost that much. It used to be advantageous to send manuscripts at a special fourth-class book rate, but that edge has all but disappeared. It's about the same rate to send materials first class, and the two-day priority mail cardboard packets aren't much more expensive, either.

Finally, be sure to keep a "submissions file," listing where you've sent your play, when you sent it, deadlines or expected response time, and the response. It's easy to forget how long ago you mailed a play or even which plays you sent to which theaters or contests.

PRODUCTIONS, AGENTS, AND CONTRACTS

Before you think about publication, you will want to get your play produced. There are numerous amateur and professional theaters that actively solicit new scripts. The best way to find out about these theaters and the kinds of scripts they want is to look at *Dramatists Source Book* or *The Playwright's Companion,* both published in New York. The *Source Book* (Theatre Communications Group) and the *Companion* (Feedback Theatre Books) each cost about $20, but they are also available in the reference rooms of most libraries. Published annually, they identify theaters willing to look at original scripts, playwriting contests, publishers and play-leasing companies, agents, arts councils, and other opportunities for playwrights. All the organizations mentioned in this chapter, as well as thousands more, are detailed in those invaluable resources. Periodicals such as *The Writer* or *Backstage* also provide information about opportunities for writers as well as extensive information about writing as a business and articles about improving your writing craft.

Theaters listed in the *Source Book* and the *Companion* range from the most prestigious New York, Chicago, and West Coast companies to professional regional theaters to college and university theaters to semi-professional, amateur, and community theaters. Requirements are as widely varied as the theaters themselves. One wants plays on Irish issues, while another wants "intellectual, poetic content." Certain theaters specialize in children's plays, in musicals, in adaptations, or in virtually any kind of drama you can imagine. A recent edition of the *Companion* listed over 600 theaters and producing organizations. Not all of them, however, produce original scripts every year. Many require submission through an agent, and some only produce work by already established authors.

Rather than sending out hundreds of scripts, you might want to explore production opportunities that are more direct and closer to home. If there are school, community, or regional theaters in your area, you should consider taking an active part. Attend the plays. Volunteer to work on the productions if you can. Get to meet the other people who are involved. Not only will you learn a lot about the theater, but the reality is that your play has a much greater chance to be read or produced by people you know than by strangers. After all, the bottom line is production. As A. R. Gurney advises, "Get the play done. Then, if you can, get it done again."

Frequently the full-scale production of a play arrives only at the end of a long and torturous process, and that process often begins with a simple informal reading. Such readings usually occur in a back room of a theater or someone's living room (often the playwright's!) with invited guests. The performers are recruited volunteers, and the reading may be the first time the actors have even seen the script.

Many theater groups also present prepared or staged readings, which afford a somewhat more formal approach to the script. In both prepared and staged readings, performers have reviewed the script with the writer or a director, discussed the characters, relationships, and themes, and rehearsed their lines. In a prepared reading, actors sit on chairs or stools with stands for their scripts. A staged reading will employ minimal set pieces and props, and the performers will move about and interact within the space—but still with scripts in hand. These readings are performed in a large room or theater for invited guests, but they are generally open to the public as well, offering an excellent chance for productive feedback for the author, both from the performers and from the audience.

The next step would be a formal production with memorized lines and proper attention to sets, props, costumes, and lights. The production might occur in a studio "black box" or on a large proscenium stage, but all the essential ingredients are present.

When you get to that point, it's time to consider agents and contracts. Many professional theaters and major producing organizations will only read scripts submitted through an agent. So how do you get an agent? Lists of agents aren't hard to find. They're available in numerous publications, including the books I've mentioned. But the fact is that agents will only be interested in your work *after* it gains some notoriety—*after* your plays have recorded successful productions or won some contests. Agents are always more receptive when you have a play that has been produced or is about to be produced. Once your play begins to achieve some positive results, agents may actually even come to you. Agents or their representatives regularly attend workshops and staged readings, and they aren't reluctant to pursue something they find interesting. After all, they're looking to represent authors and plays that they think will gain success. But if you're an unknown writer, you can't expect an agent or anyone else to do the difficult spadework of marketing your plays. In the early stages of your career as a writer, only you can provide the dedicated labor you need.

The opportunity may arise when you require a formal contract between you as the playwright and the theater producing your play. At that time an agent or a lawyer who specializes in the arts can assist you. You can also purchase sample contracts through The Dramatists Guild in New York, a professional association of playwrights, composers, and lyricists. Such contracts specify the rights and responsibilities of the author and of the producing organization. They stipulate procedures in case of disagreements between the parties, and they suggest typical financial guidelines.

Sometimes it takes years for plays to progress from one stage of development to another. I recall attending a small regional production of Tom Ziegler's off-Broadway hit *Grace & Glorie* several years before its successful New York run. Tony Kushner's *Angels in America* underwent a similarly lengthy development before it achieved its award-winning status. That pattern has become typical for most successful new scripts. This year's new hit has probably been at least five or six years in maturation.

Often writers and local producers decide to simply produce their plays themselves. One of my former students, a high school teacher, grew weary of the annual search for the right play to fit his changing student populations, so he began writing his own. His original productions met with enthusiastic receptions. Other drama coaches wanted to produce the plays. As a result, several of his scripts are now handled by a major play-leasing company, and one of his plays ranks among the most produced one acts in the country!

Another former student founded a theater in the middle of West Virginia. She wanted her company to reflect and comment on the people, events, and heritage of that area, so she and fellow troupers wrote and produced their own scripts. With dynamic individuals and lots of marketing elbow grease, the old "Let's put on a play!" mentality can work wonders!

CONTESTS

Another approach to production is to enter your play in appropriate contests. You can be assured of a careful reading, and in most cases an actual winner or winners are named and receive monetary rewards.

Requirements for the contests vary. Some contests exclude musicals while others are *only* for musicals. There are several contests for children's scripts. Some accept only full-length plays, others want only one acts. There's a contest for ten-minute plays and another for one-minute plays!

In many contests entrants must be from a particular state, area, or region. In a few the author must be of a specific race, gender, age, or religion. Contests often include rules about the sets or numbers of characters that should be used. Most theaters today are looking for scripts with minimal scenic requirements such as one set and a handful of characters. In addition, many theaters are especially looking for scripts with good female roles.

There are contests for translations of plays into English, plays in other languages, and plays adapted from other sources. Of course, whenever authors are dealing with a translation or an adaptation, they must be careful to secure the appropriate rights to the original material.

Sometimes requirements for subject matter are specified. There are contests that seek scripts about the environment, world peace, Afro-American experiences, Jewish values, women's issues, Hispanic-American concerns, gay rights, Christian themes, and deaf culture. Numerous contests focus on material about specific states or regions: plays about Hawaii, the "Republic of Texas," Appalachia, the South, or rural America.

Just as requirements vary from contest to contest, so, too, do your odds of winning. The most prestigious contests with the most substantial prizes naturally attract more applicants. Some contests generate thousands of entrants, while others receive substantially fewer. Those figures are usually included in *The Playwright's Companion* and *Dramatists Source Book.*

A recent edition of *The Playwright's Companion* identified over 200 play contests and listed various criteria such as deadlines and contacts. A few of the contests charged small entrance fees. In most cases prizes ranged from $100 to $5000. In almost all cases the winning scripts were assured of a staged reading or a full production. Theaters that produce prize-winning plays often bring the playwright in for the production and sometimes for the rehearsal process as well.

RETREATS, RESIDENCIES, SEMINARS, AND THE WEB

Retreats, residencies, and seminars afford excellent opportunities for gaining exposure, developing your script, and networking for possible productions. One problem with submitting plays to theaters or to contests is the lack of feedback. Because so many scripts are sent, playwrights seldom receive any comments on their work. Playwriting workshops and seminars can generate valuable responses.

Different organizations boast different workshop formats. Several provide openings for eight to ten playwrights to spend anywhere from two weeks to two months writing, discussing, and staging their works. The sites are usually secluded, hence the idea of a "retreat." The O'Neill National Playwrights Conference in New York, The Sundance Playwrights Laboratory in Utah, the Mount Sequoyah New Play Retreat in Arkansas, the New Harmony Project in Indiana, and the Shenandoah Playwrights Retreat in Virginia all operate in that fashion.

Other programs admit just one playwright at a time, and those can last from a week to a year. Such solo programs are usually referred to as "residencies" because the writer is in residence for a period of time. Some residency programs are located in secluded sites, but others are in metropolitan areas. Retreats and residencies generally furnish board and housing for the writers. Although some organizations charge modest fees, at the other end of the spectrum, some residencies also provide attractive stipends in addition to room and board.

The primary purpose of such retreats and residencies is to give writers time to write. Retreats with other writers have the additional benefit of discussion and feedback, which encourages a lively atmosphere of artists engaging with other artists on equal terms.

Seminars can also be helpful to your growth as a playwright. Like retreats, seminars typically gather a few writers at a secluded location for a concentrated period of time—usually a week or two. In major theatrical centers such as New York, London, Chicago, or Los Angeles, seminars may run over a period of months, meeting once or twice a week.

Seminars include an instructional component, with professionals engaged as teachers and mentors. Participants pay fees—usually several hundred dollars per week—for the instruction, the feedback, and the networking opportunities.

And, speaking of "networks," a new source of information and inspiration that has just recently emerged is the Internet or the World Wide Web. You can find scripts, interviews with writers, helpful tips, writing tutorials, and basic information on subjects ranging from copyrights and contests to theaters and professional writers' organizations. A few of the more intriguing and helpful sites for dramatic writers are:

> The Playwriting Seminars
> http://www.vcu.edu/artweb/playwriting/seminar.html
> Disseminates Richard Toscan's instruction, which is valuable for all levels of writers.

> Essays on the Craft of Dramatic Writing
> http://www.teleport.com/~bjscript/index.htm
> Focuses on worthwhile explorations regarding fundamental issues of storytelling.

> E-Script: The Internet's Scriptwriting Workshop
> http://www.singlelane.com/escript/
> Supplies online workshops in dramatic writing for theater, film, and television.

> Playbill on Line
> http://www1.playbill.com/playbill/
> Gives loads of news, features, and links to other sites in all categories of theater.

> Screenwriters & Playwrights Home Page
> http://www.teleport.com/~cdeemer/scrwriter.html
> Dispenses tips from professional writers, discussion group chats, and a slew of practical information.

> The Dramatic Exchange
> http://www.dramex.org/
> Specializes in archiving and distributing scripts.

Screenwriters Online
http://www.screenwriter.com/main html
Contains information, interviews with professionals in the entertainment
business, and commercial services.

Highlights for Screenwriters
http://www.teleport.com/~cdccmer/scr-highlight.html
Dispenses sound, basic tutorial instruction for screenwriters.

Screenwriters Resource Center
http://www.screenwriting.com
Provides informative resources and links to numerous other helpful sites.

Screenwriting for Motion Pictures & Television
http://www.sellingtohollywood.com/link.html
Features an interesting question and answer format for a variety of catego-
ries as well as basic information and links to additional sites.

The Screenwriter's Utopia
http://www.screenwritersutopia.com/
Includes entertainment industry news and interviews with professional
writers.

PLAY-LEASING COMPANIES AND PUBLISHERS

Just as inexperienced writers think that the most important parts of
plays are the lines of dialogue, so many inexperienced playwrights believe
the most important thing to do with a play is to get it published. Actually,
there is little point in submitting your play directly to a publisher. Harcourt
Brace, Dutton, Random House, or Faber and Faber will likely only be
interested in publishing your play for its literary value *after* it has achieved
success as a theatrical production. From a financial perspective, playwrights
derive far more income from production royalties than from sales of the
scripts. For those writers looking for publication outlets, however, there are
some magazines that regularly publish scripts, including *Plays, The Dramatic
Magazine for Young People; Dramatics Magazine,* whose readership consists
primarily of high school students and teachers; or *Theater Magazine,* which
is widely read by theater professionals.

The majority of published scripts are put out by play-leasing
companies rather than traditional book publishers. Play leasing companies
such as Samuel French or Dramatists Play Service serve as a liaison between playwrights, the scripts they represent, and people who want to
produce the play. If a play-leasing company likes your play and agrees to
represent it, they will publish it in an acting edition and include it in their
catalogue of available plays. If someone wants to stage the play, the company handles all correspondence, scheduling, details of contracts, and
collection of royalties. Then, usually once or twice a year, they send to you,
the author, a report detailing when and where productions were mounted,
by whom, the number of performances, the number of scripts sold, and a
check for your portion of the royalties. For performing all those services,

the company keeps a percentage of the sales and royalty fees to cover their expenses and ensure their profit.

Just like traditional literary publishing houses (which, by the way, usually have nothing to do with *production* rights to plays), the major play-leasing companies are normally interested in putting out a production version of a script and arranging for production rights only *after* a play has received a successful run. In the words of a Dramatists Play Service representative, "We seldom find a market for a play that does not have a commercially successful production to its credit."

However, virtually all of the play-leasing companies will read unsolicited, unproduced plays, and sometimes they will offer a contract for an unknown work. In a recent year, out of 1500 submitted scripts, Samuel French published 60. Response can take anywhere from two months to a year or more. Be sure to consider the smaller companies such as I. E. Clark, Pioneer Drama Service, Broadway Play Publishing, and Contemporary Drama Service along with the majors—Samuel French, Dramatists, and Dramatic Publishing.

WRITING YOUR SECOND PLAY

The title of this book is *Writing Your First Play,* but there's no reason to stop there. After you've written your first play and revised it as well as you can, start writing your second play! I've known too many writers who spent their whole lives constantly revising their one play. As Terrence McNally has advised, "Just go on and write the next one." If you are fortunate enough to have one of your plays produced, the production process will provide you ample opportunity and reason for revision.

Getting a play produced can be a long and arduous process. Sometimes the timeliness of a current issue evaporates before the play is ever performed. Frequently plays must be rewritten again and again to satisfy the needs of producing organizations. Tom Ziegler, author of *Grace & Glorie,* has written another play that is currently in planning for a New York premiere. I first saw this second play at a staged reading several years ago. A year after that I directed it with a full university staging, and only now will it receive a major New York production. Still, despite all the hurdles, an undeniable and unforgettable magic occurs when other artists—actors, directors, and designers—take your script and add their talents to it to bring forth a living, artistic creation.

Appendix

· · · · · · · ·

Classroom Procedures

Some readers of this book may want to use these exercises in a classroom setting. The sequence of exercises I've outlined has been used successfully by instructors at several colleges and universities, at the Virginia Summer program for the Gifted and Talented, as well as by English and drama teachers to develop a unit on playwriting for their classes.

Because I believe that *how* something is done shares importance with *what* is done, I've suggested some ways of proceeding. Because I know that teachers and their classes vary, I offer these suggestions as just that: suggestions.

1. The author or other participants should read the play to the group.

There are a variety of ways to disseminate to the rest of the class a scene written by a student. Some instructors like to make copies of the piece, pass them out, and have everyone read the scenes outside of class time. That may be necessary if there is insufficient class time, but you lose the sound of the scene, which I think is a large sacrifice.

Many instructors like to distribute the parts in a scene to the students in class, making as many copies as there are characters in the scene plus one for stage directions. That way, different people read different parts, more students are actively involved, and the author can hear what it sounds like with other people acting the words. This technique also offers the added dimension of what the "performers" reading the role contribute to it by their delivery of lines.

Another way is simply to have students read their own pieces. That gets as close to the author's conception as possible because you hear a line the way the author wants it read. That approach makes the connection from the author's mind to the work we hear as direct as possible, and it is the approach most often used at playwriting retreats when authors present their material for the first time.

One additional advantage of having authors read their pieces themselves is that it encourages the writers to take responsibility for what they're writing. In professional writing situations, authors must be able to "pitch" their material; as we saw in Chapter 10, they must be their own best salespeople.

I have alternated with different approaches. I used to like authors to read their own material, but more recently, with ever larger classes, I've liked the fact that assigning parts to other participants more actively involves the whole group.

> 2. As many students as practical should comment on each work that is read.

I'll divide a group of more than a dozen students in half, and ask at least half the people to comment on a scene. With fewer than that, everyone should comment on every piece. This practice generates a climate in which each person exercises analytical skills on each play, and every student feels somehow a part of every other student's play. Instructors may also find it helpful for students to keep a notebook or journal of their comments, questions, and reactions to every work presented.

Every student will not be an exceptional playwright. But every student can gain an understanding of the process and the problems of writing plays. Every student can attain an appreciation for the difficulties of structuring plot, developing character, instigating conflict, and devising dialogue. Students learn from hearing, analyzing, and commenting as well as from their own writing.

For those reasons, I regard the commenting time as essential. At first students may be reluctant to critique the work of others. They may feel they don't know how to do it, what to look for, or what the teacher wants. They may need guidelines, such as: "Start out by saying what you liked about the piece, what caught your attention or interested you. You can also mention things you didn't like or didn't understand." You should steer students away from rewriting other students' work, which they sometimes try to do by suggesting changes or indicating how they would solve a problem or make something better.

> 3. The writer should refrain from editorial comment until all remarks are concluded.

I've often run into students who are great talkers and want to *tell* us about their scenes. That must be avoided. No explaining before the scene starts. No justifying after it's over. The dramatic material must, like the cheese, stand alone. Auditors should be responding to the scene and only to the scene, not to the author's commentary on the scene. This is precisely the regimen employed for new play development at New York's Circle Repertory Company. Lanford Wilson explains it this way: "The writer's not allowed to answer any questions or you start talking about what you intended. What you are trying to understand is what the audience got, and your intentions be damned."

After all other remarks and questions are completed, the author may be given an opportunity to comment. Authors may legitimately want to know if something they were trying to accomplish succeeded. Or they may ask for suggestions in solving a particularly troubling problem. They should not, however, use the time simply as a self-defense session.

Bibliography

Anderson, Maxwell. *The Essence of Tragedy*. New York: Russell and Russell, 1970.

Andrews, Richard. *Writing a Musical*. New York: Parkwest Publications, 1997.

Archer, William. *Play-making*. Boston: Small, Maynard and Company, 1912.

Armer, Alan A. *Writing the Screenplay*. Belmont, Calif.: Wadsworth, 1988.

Baker, George Pierce. *Dramatic Technique*. New York: De Capo Press, 1971.

Berman, Robert. *Fade In: The Screenwriting Process*. Studio City, Calif.: Michael Wiese Productions, 1997.

Blackr, Irwin R. *The Elements of Screenwriting*. New York: Macmillan, 1988.

Brady, Ben. *The Keys to Writing for Television and Films*. Dubuque, Iowa: Kendall/Hunt, 1982.

Brady, Ben and Lance Lee. *The Understructure for Writing for Film and Television*. Austin: University of Texas Press, 1988

Brady, John, ed. *The Craft of the Screenwriter: Interviews with Six Celebrated Screenwriters*. New York: Simon & Schuster, 1982.

Bronfeld, Stewart. *Writing for Film and Television*. Englewood Cliffs, N. J.: Prentice-Hall, 1981.

Brown, Cary. *The Screenwriter's Companion*. Beverly Hills, Calif.: Silman-James Press, 1994.

Busfield, Roger M., Jr. *The Playwright's Art*. Westport, Conn.: Greenwood Press, 1971.

Callan, K. *The Script Is Finished, Now What Do I Do?* New York: Sweden Press, 1997.

Cassady, Marsh. *Characters In Action: Playwriting the Easy Way*. Colorado Springs: Meriwether Publishing , 1995.

Catron, Louis E. *The Elements of Playwriting*. Old Tappen, N. J.: Macmillan, 1993.

——. *Writing, Producing, and Selling Your Play*. Prospect Heights, Ill.: Waveland Press, 1990.

Checkov, Michael. *To the Director and Playwright*. Compiled by Charles Leonard. Westport, Conn.: Greenwood Press, 1977.

Cohen, Edward M. *Working on a New Play*. New York: Limelight Editions, 1995.

Cole, Toby, ed. *Playwrights on Playwriting*. New York: Hill and Wang, 1961.

Dmytryk, Edward. *On Screen Writing*. Boston: Focal Press, 1985.

Downs, William Missouri and Lou Anne Wright. *Playwriting: From Formula to Form*. Ft. Worth, Tex.: Harcourt Brace College Publishers, 1998.

Dyas, Ronald D. *Screenwriting for Television and Film*. Madison, Wis.: Brown & Benchmark, 1993.

Edmonds, Robert. *Scriptwriting for the Audio-Visual Media*. New York: Teachers College Press, 1984.

Egri, Lajos. *How to Write a Play*. New York: Simon and Schuster, 1942.

——. *The Art of Dramatic Writing*. New York: Simon and Schuster, 1972.

Engel, Joel. *Screenwriters on Screenwriting.* New York: Hyperion, 1995.

Fann, Ernest L. *How to Write a Screenplay.* Los Angeles: David-Kristy, 1988.

Field, Syd. *Screenplay: The Foundations of Screenwriting.* New York: Dell, 1984.

——. *The Screenwriter's Workbook.* New York: Dell, 1988.

Finch, Robert. *How to Write A Play.* New York: Greenberg Press, 1948.

Frankel, Aaron. *Writing the Broadway Musical.* New York: Drama Book Specialists, 1977.

Frensham, Raymond G. *Teach Yourself Screenwriting.* Lincolnwood, Ill.: NTC Publishing Group, 1996.

Freytag, Gustav. *The Technique of the Drama.* New York: B. Blom, 1968.

Friedman, Robert. *Playwright Power.* Lanham, Md.: University Press of America, 1996.

Frome, Shelly. *Playwriting: A Complete Guide to Creating Theater.* Jefferson, N.C.: McFarland & Company, 1990.

Froug, William. *Zen and the Art of Screenwriting.* Beverly Hills, Calif.: Silman-James Press, 1996.

Funke, Lewis, ed. *Playwrights Talk About Writing.* Chicago: Dramatic Publishing, 1975.

Geller, Stephen. *Screenwriting: A Method.* New York: Bantam Books, 1984.

George, Kathleen E. *Playwriting: The First Workplace.* Boston: Focal Press, 1994.

Gibson, William. *Shakespeare's Game.* New York: Atheneum, 1978.

Gillis, Joseph. *The Screenwriter's Guide.* New York: New York Zoetrope, 1987.

Giustini, Roland. *The Filmscript: A Writer's Guide.* Englewood Cliffs, N. J.: Prentice-Hall, 1980.

Goldman, William. *Adventures in the Screen Trade: A Personal View of Hollywood and Screenwriting.* New York: Warner Books, 1989.

Goodman, Evelyn. *Writing Television and Motion Picture Scripts That Sell.* Chicago: Contemporary Books, 1982.

Granville-Barker, Harley. *On Dramatic Method.* London: Sidgwick and Jackson, 1931.

Grebanier, Bernard. *Playwriting.* New York: Barnes & Noble, 1961.

Griffiths, Stuart. *How Plays Are Made.* Portsmouth, N. H.: Heinemann, 1988.

Gum, Lori. *How to Write a Play.* Syracuse, N.Y.: New Readers Press, 1991.

Haag, Judith H. and Cole, Hollis R., Jr. *The Complete Guide to Standard Script Formats.* Hollywood: CMC Publishing, 1980.

Hatcher, Jeffrey. *The Art and Craft of Playwriting.* Cincinnati: Betterway Books, 1996.

Hatton, Thomas J. *Playwriting for Amateurs.* Downer's Grove, Ill.: Meriwether, 1981.

Hauge, Michael. *Writing Screenplays That Sell.* New York: HarperCollins, 1991.

Herman, Lewis. *A Practical Manual of Screen Playwriting for Theater and Television Films.* Cleveland: World Publishing, 1966.

Howard, David. *Tools of Screenwriting.* New York: St. Martin's Press, 1995.

Hull, Raymond. *How to Write a Play.* Cincinnati: Writer's Digest Books, 1983.

——. *Profitable Playwriting.* New York: Funk & Wagnalls, 1968.

Hunter, Lew. *Lew Hunter's Screenwriting.* New York: Bradley Publishing Group, 1994.

Josefsberg, Milt. *Comedy Writing for Television and Hollywood.* New York: Perennial Library, 1987.

Kahn, David. *A Director's Approach to New Play Development.* Carbondale, Ill.: Southern Illinois University Press, 1995.

Kerr, Walter. *How Not to Write a Play.* New York: Simon and Schuster, 1955.

King, Viki. *How to Write a Movie in 21 Days.* New York: Harper & Row, 1988.

Kline, Peter. *Playwriting*. New York: R. Rosen Press, 1970

Kohn, Philip C. *Speaking on Stage: Interviews with Contemporary American Playwrights* Tuscaloosa. University of Alabama Press, 1996.

Korty, Carol. *Writing Your Own Plays*. New York: Scribner, 1986.

Langer, Lawrence. *The Play's the Thing*. Boston: The Writer, 1960.

Lawson, John Howard. *Theory and Technique of Playwriting and Screenwriting*. New York: Garland Publishing, 1985.

Lee, Donna. *Magic Methods of Screenwriting*. New York: McGraw-Hill, 1995.

Longman, Stanley Vincent. *Composing Drama for Stage and Screen*. Boston: Allyn and Bacon, 1986.

MacGowan, Kenneth. *A Primer of Playwriting*. Westport, Conn.: Greenwood Press, 1981.

Maloney, Martin and Paul Max Rubenstein. *Writing for the Media*. Englewood Cliffs, N.J.: Prentice Hall, 1980.

Marion, Frances. *How to Write and Sell Film Stories*. New York: Garland Publishing, 1978.

Matthews, Brander, ed. *Papers on Playmaking*. North Stratford, N.H.: Ayer Company, 1977.

Mayfield, William F. *Playwriting for Black Theatre*. Pittsburgh: William F. Mayfield, 1985.

McLaughlin, Buzz. *The Playwright's Process*. New York: Back Stage Books, 1997.

Merserve, Mollie Ann, comp. *The Playwright's Companion: A Submission Guide to Theatres and Contests in the U. S. A.* New York: Feedback Theatre Books, published annually.

Miller, J. William. *Modern Playwrights at Work*. New York: Samuel French, 1968.

Miller, William C. *Screenwriting for Narrative Film and Television*. New York: Hastings House, 1980.

Muth, Marcia. *How to Write and Sell Your Plays*. Santa Fe, N.M.: Sunstone Press, 1974.

Nash, Constance and Virginia Oakey. *The Screenwriter's Handbook*. New York: HarperCollins, 1978.

Nigli, Josefina. *New Pointers on Playwriting*. Boston: The Writer, 1967.

Nolan, Paul T. *Writing the One-Act Play for the Amateur Stage*. Denver: Pioneer Drama Service, 1977.

Norton, James H. and Francis Gretton. *Writing Incredibly Short Plays, Poems, and Stories*. New York: Harcourt Brace Jovanovich, 1972.

Packard, William. *The Art of the Playwright*. New York: Paragon House, 1987.

——. *The Art of Screenwriting*. New York: Thunder Mouth Press, 1997.

Pike, Frank and Thomas G. Dunn. *The Playwright's Handbook*. New York: Penguin Books, 1996.

Polsky, Milton E. *You Can Write a Play*. New York: Rosen Publishing Group, 1983.

Polti, Georges. *The Thirty-Six Dramatic Situations*. Boston: The Writer, 1954.

Rilla, Wolf. *The Writer and the Screen*. New York: William Morrow, 1974.

Root, Wells. *Writing the Script*. New York: Holt, Rinehart, and Winston, 1980.

Rowe, Kenneth T. *Write That Play*. New York: Minerva Press, 1968.

Saulter, Carl. *How to Sell Your Screenplay*. New York: New Chapter Press, 1988.

Savran, David, ed. *In Their Own Words: Contemporary American Playwrights*. New York: Theatre Communications Group, 1988.

Seger, Linda. *Making a Good Script Great*. New York: Samuel French, 1994.

Shamas, Laura. *Playwriting for Theatre, Film, and Television*. Cincinnati: Betterway Books, 1991.

Shanks, Bob. *The Primal Screen*. New York: Fawcett, 1987.

Smiley, Sam. *Playwriting: The Structure of Action*. Englewood Cliffs, N. J.: Prentice-Hall, 1971.

Sova, Kathy and Wendy Weiner, comps. *Dramatists Source Book*. New York: Theatre Communications Group, published annually.

Spira, Robert. *Playwrighting*. Ashland, Ore: Quartz Press, 1991.

Straczynski, J. Michael. *The Complete Book of Scriptwriting*. Cincinnati: Betterway Books, 1996.

Swain, Dwight V. *Film Scriptwriting*. Boston: Focal Press, 1988.

Sweet, Jeff. *The Dramatist's Tool Kit*. Portsmouth, N. H.: Heinemann, 1993.

Trottier, David. *The Screenwriter's Bible*. Salt Lake City, Utah.: Clearstream, 1995.

Vale, Eugene. *The Technique of Screen and Television Writing*. New York: Simon and Schuster, 1986.

Van Druten, John. *Playwright at Work*. New York: Harper & Brothers, 1953.

Wager, Walter, ed. *The Playwrights Speak*. New York: Dell, 1968.

Walter, Richard. *Screenwriting*. New York: NAL Dutton, 1988.

Weales, Gerald Clifford. *A Play and Its Parts*. New York: Basic Books, 1964.

Whitcomb, Cynthia. *Selling Your Screenplay*. New York: Crown, 1988.

Willis, Edgar E. and Camilee D'Arienzo. *Writing Scripts for Television, Radio, and Film*. New York: Holt, Rinehart, and Winston, 1981.

Wolff, Jurgen and Kerry Cox. *Successful Scriptwriting*. Cincinnati: Betterway Books, 1988.

Wolff, Jurgen. *Top Secrets: Screenwriting*. Los Angeles: Lone Eagle Publishing, 1993.

Wright, Michael. *Playwriting-in-Process*. New York: Heinemann, 1997.

Index

10/99 - 5/07 - 23

DISCARD/SOLD
FRIENDS MLS